THE DRIVER IN THE DRIVERLESS CAR

VIVEK WADHWA WITH ALEX SALKEVER

THE DRIVER IN THE DRIVERLESS CAR

HOW OUR TECHNOLOGY CHOICES WILL CREATE THE FUTURE

BK®

Berrett–Koehler Publishers, Inc.
a BK Business book

Berrett-Koehler Publishers, Inc.
1333 Broadway, Suite 1000
Oakland, CA 94612-1921
Tel: (510) 817-2277; Fax: (510) 817-2278
www.bkconnection.com

Ordering Information

Quantity sales. Special discounts are available on quantity purchases by corporations, associations, and others. For details, contact the "Special Sales Department" at the Berrett-Koehler address above.

Individual sales. Berrett-Koehler publications are available through most bookstores. They can also be ordered directly from Berrett-Koehler: Tel: (800) 929-2929; Fax: (802) 864-7626; www.bkconnection.com.

Orders for college textbook/course adoption use. Please contact Berrett-Koehler: Tel: (800) 929-2929; Fax: (802) 864-7626.

Orders by U.S. trade bookstores and wholesalers. Please contact Ingram Publisher Services: Tel: (800) 509-4887; Fax: (800) 838-1149; E-mail: customer.service@ingrampublisherservices.com; or visit www.ingrampublisherservices.com/Ordering for details about electronic ordering.

Printed in the United States of America

Berrett-Koehler books are printed on long-lasting acid-free paper. When it is available, we choose paper that has been manufactured by environmentally responsible processes. These may include using trees grown in sustainable forests, incorporating recycled paper, minimizing chlorine in bleaching, or recycling the energy produced at the paper mill.

Library of Congress Cataloging-in-Publication Data

Names: Wadwha, Vivek, author. | Salkever, Alex, author.
Title: The driver in the driverless car : how our technology choices will create the future / Vivek Wadwha and Alex Salkever.
Description: First edition. | Oakland, CA : Berrett-Koehler Publishers, Inc., [2017] | "A BK Business book."
Identifiers: LCCN 2016052693 | ISBN 9781626569713 (hardcover : acid-free paper)
Subjects: LCSH: Technology—Social aspects. | Technological forecasting.
Classification: LCC HM846.W33 2017 | DDC303.48/3—dc23
LC record available at https://lccn.loc.gov/2016052693

First Edition
22 21 20 19 18 17 || 10 9 8 7 6 5 4 3

Produced by BookMatters; copyedited by Tanya Grove; proofread by Janet Reed Blake; indexed by Leonard Rosenbaum; cover by Rob Johnson, Toprotype, Inc.

CONTENTS

PART THREE
What Are the Risks and the Rewards?

PART FOUR
Does the Technology Foster Autonomy or Dependency?

PREFACE

Not long ago, I was very pessimistic about the future. I was worried about hunger and poverty, disease, overpopulation. I believed that the world would run out of clean water and energy and that we would be fighting world wars over scarce resources.

Today, I talk about this being the greatest period in history, when we will solve the grand challenges of humanity and enter an era of enlightenment and exploration such as we saw in my favorite TV series, *Star Trek*. Yes, I grew up dreaming of tricorders, replicators, and androids and wanting to be an astronaut so that I could join Starfleet Academy. Didn't all the people from my generation, of the '60s?

At Stanford, Duke, and Singularity universities, and now at Carnegie Mellon, I have spent the past six years researching the advances in technology that are finally making science fiction a reality. It truly is amazing what is possible, as I will explain in this book. But I have come to realize that reaching Utopia will take vigilance and effort: like the course of a game of snakes and ladders, our path is strewn with hazards.

My research has made me acutely aware of the dangers

in advanced technologies. These are moving faster than people can absorb change—and offer both unprecedented rewards and unpredictable hazards.

As a society, we can make amazing things happen; and the more we understand, the better our decision making will be—and the greater the odds that we head toward *Star Trek*. Today's technology changes are happening so quickly and are so overwhelming that all of us—including technologists—can benefit from access to new tools for considering and managing them. I wrote this book with the help of my good friend and writing guru, Alex Salkever, in order to provide such tools, because I believe in the power of choice and the greater judgment of involved citizens. We hope that it will help you deal with the challenges that new technologies raise now and in the future.

INTRODUCTION

It is a warm autumn morning, and I am walking through downtown Mountain View, California, when I see it. A small vehicle that looks like a cross between a golf cart and a Jetsonesque bubble-topped spaceship glides to a stop at an intersection. Someone is sitting in the passenger seat, but no one seems to be sitting in the driver seat. How odd, I think. And then I realize I am looking at a Google car. The technology giant is headquartered in Mountain View, and the company is road-testing its diminutive autonomous cars there.

This is my first encounter with a fully autonomous vehicle on a public road in an unstructured setting.

The Google car waits patiently as a pedestrian passes in front of it. Another car across the intersection signals a left-hand turn, but the Google car has the right of way. The automated vehicle takes the initiative and smoothly accelerates through the intersection. The passenger, I notice, appears preternaturally calm.

I am both amazed and unsettled. I have heard from friends and colleagues that my reaction is not uncommon. A driverless car can challenge many assumptions about human superiority to machines.

Though I live in Silicon Valley, the reality of a driverless car is one of the most startling manifestations of the future unknowns we all face in this age of rapid technology development. Learning to drive is a rite of passage for people in materially rich nations (and becoming so in the rest of the world): a symbol of freedom, of power, and of the agency of adulthood, a parable of how brains can overcome physical limitations to expand the boundaries of what is physically possible. The act of driving a car is one that, until very recently, seemed a problem only the human brain could solve.

Driving is a combination of continuous mental risk assessment, sensory awareness, and judgment, all adapting to extremely variable surrounding conditions. Not long ago, the task seemed too complicated for robots to handle. Now, robots can drive with greater skill than humans—at least on the highways. Soon the public conversation will be about whether humans should be allowed to take control of the wheel at all.

This paradigm shift will not be without costs or controversies. For sure, widespread adoption of autonomous vehicles will eliminate the jobs of the millions of Americans whose living comes of driving cars, trucks, and buses (and eventually all those who pilot planes and ships). We will begin sharing our cars, in a logical extension of Uber and Lyft. But how will we handle the inevitable software faults that result in human casualties? And how will we program the machines to make the right decisions when faced with impossible choices—such as whether an autonomous car should drive off a cliff to spare a busload of children at the cost of killing the car's human passenger?

I was surprised, upon my first sight of a Google car on the street, at how mixed my emotions were. I've come to realize that this emotional admixture reflects the countercurrents that the bow waves of these technologies are rocking all of us with: trends toward efficiency, instantaneity, networking, accessibility, and multiple simultaneous media streams, with consequences in unemployment, cognitive and social inadequacy, isolation, distraction, and cognitive and emotional overload.

Once, technology was a discrete business dominated by business systems and some cool gadgets. Slowly but surely, though, it crept into more corners of our lives; today, that creep has become a headlong rush. Technology is taking over everything: every part of our lives, every part of society, every waking moment of every day. Increasingly pervasive data networks and connected devices are enabling rapid communication and processing of information, ushering in unprecedented shifts—in everything from biology, energy, and media to politics, food, and transportation—that are redefining our future. Naturally we're uneasy; we should be. The majority of us, and our environment, may receive only the backlash of technologies chiefly designed to benefit a few. We need to feel a sense of control over our own lives; and that necessitates actually having some.

The perfect metaphor for this uneasy feeling is the Google car. We welcome a better future, but we worry about the loss of control, of pieces of our identity, and most importantly of freedom. What are we yielding to technology? How can we decide whether technological innovation that alters our lives is worth the sacrifice?

The noted science-fiction writer William Gibson, a favorite of hackers and techies, said in a 1999 radio interview (though apparently not for the first time): "The future is already here; it's just not very evenly distributed."[1] Nearly two decades later—though the potential now exists for most of us, including the very poor, to participate in informed decision making as to its distribution and even as to bans on use of certain technologies—Gibson's observation remains valid.

I make my living thinking about the future and discussing it with others, and am privileged to live in what to most is the future. I drive an amazing Tesla Model S electric vehicle. My house, in Menlo Park, close to Stanford University, is a Passive House, extracting virtually no electricity from the grid and expending minimal energy on heating or cooling. My iPhone is cradled with electronic sensors that I can place against my chest to generate a detailed electrocardiogram to send to my doctors, from anywhere on Earth.*

Many of the entrepreneurs and researchers I talk with about breakthrough technologies, such as artificial

*I have a history of heart trouble, including a life-threatening heart attack; my ability to communicate with my doctors in seconds instead of hours makes my life both safer and easier, and gives me the confidence to go hiking up mountains and to travel the world giving talks.

intelligence and synthetic biology, are building a better future at a breakneck pace. One team built a fully functional surgical-glove prototype to deliver tactile guidance for doctors during examinations—in three weeks. Another team's visualization software, which can tell farmers the health of their crops using images from off-the-shelf drone-flying video cameras, took four weeks to build.

The distant future, then, is no longer distant. Rather, the institutions we expect to gauge and perhaps forestall new technologies' hazards, to distribute their benefits, and to help us understand and incorporate them are drowning in a sea of change as the pace of technological change outstrips them.

The shifts and the resulting massive ripple effects will, if we choose to let them, change the way we live, how long we live, and the very nature of being human. Even if my futuristic life sounds unreal, its current state is something we may laugh at within a decade as a primitive existence—because our technologists now have the tools to enable the greatest alteration of our experience of life that we will have seen since the dawn of humankind. As in all other manifest shifts—from the use of fire to the rise of agriculture and the development of sailing vessels, internal-combustion engines, and computing—this one will arise from breathtaking advances in technology. It is far larger, though, is happening far faster, and may be far more stressful to those living through this new epoch. Inability to understand it will make our lives and the world seem even more out of control.

In the next few chapters, I will take you into this future, discussing some of the technologies that are advancing at an exponential pace and illustrating what they make possible. You will see how excited I am about their potential—and how worried, at the same time, about the risks that they create.

Broadly speaking, we will, jointly, choose one of two possible futures. The first is a utopian *Star Trek* future in which our wants and needs are met, in which we focus our lives on the attainment of knowledge and betterment of humankind. The other is a *Mad Max* dystopia: a frightening and alienating future, in which civilization destroys itself.

These are both worlds of science fiction created by Hollywood, but either could come true. We are already capable of creating a world of tricorders, replicators, remarkable transportation technologies, general wellness, and an abundance of food, water, and energy. On the other hand, we are capable too now of ushering in a jobless economy, the end of all privacy, invasive medical-record keeping, eugenics, and an ever worsening spiral of economic inequality: conditions that could create an unstable, Orwellian, or violent future that might undermine the very technology-driven progress that we so eagerly anticipate. And we know that it is possible to inadvertently unwind civilization's progress. It is precisely what Europe did when, after the Roman Empire, humanity slid into the Dark Ages, a period during which significant chunks of knowledge and technology that the Romans had hard won through trial and error disappeared from the face of the Earth. To unwind our own civilization's amazing progress will require merely cataclysmic instability.

It is the choices we all make that will determine the outcome. Technology will surely create upheaval and destroy industries and jobs. It will change our lives for better and for worse simultaneously. But we can reach *Star Trek* if we can share the prosperity we are creating and soften its negative impacts, ensure that the benefits outweigh the risks, and gain greater autonomy rather than becoming dependent on technology.

You will see that there is no black and white; the same technologies that can be used for good can be used for evil in a continuum limited only by the choices we make jointly. All of us have a role in deciding where the lines should be drawn.

At the end of the day, you will realize that I am an optimistic at heart. I sincerely believe that we will all learn, evolve, and come together as a species and do amazing things.

With that, let the journey begin.

PART ONE

The Here and Now

◆ 1 ◆

A Bitter Taste of Dystopia

The 2016 presidential campaign made everybody angry. Liberal Bernie Sanders supporters were angry at allegedly racist Republicans and a political system they perceived as being for sale, a big beneficiary being Hillary Clinton. Conservative Donald Trump supporters were furious at the decay and decline of America, and at how politicians on both sides of the aisle had abandoned them and left a trail of broken promises. Hillary Clinton supporters fumed at how the mainstream media had failed to hold Trump accountable for lewd behavior verging on sexual assault—and worse.

The same rage against the system showed up in Britain, where a majority of citizens primarily living outside of prosperous London voted to take England out of the European Union. In Germany, a right-wing party espousing a virulent brand of xenophobia gained critical seats in the Bundestag. And around the world in prosperous countries, anger simmered, stoked by a sense of loss and by raging income inequality. In the United States, real incomes have been falling for decades. Yet in the shining towers of finance and on kombucha-decked tech campuses for glittering growth engines such as Google and Apple, the gilded class of technology employees and Wall Street types continue to enjoy tremendous economic gains.

The roots of the rage are, in my opinion, traceable to the feelings of powerlessness that have been building since the incursion into our lives of the microprocessor and the computer. At first, we greeted computers with a sense of wonder. Simple things such as spreadsheets, word processors, and arcade-quality video games could be run on tiny boxes in our living rooms!

The technology wove deeper into our lives. E-mail replaced paper mail. Generations of Americans will never write a full letter by hand. Social networks reinstated lost connections and spread good tidings. Discussions flourished. Maps went from the glovebox to the smart phone, and then replaced our own sense of navigation with computer-generated GPS turn-by-turn guidance so prescient that neither I nor most of my friends can remember the last time we printed out directions to a party or a restaurant.

As the new electronics systems grew smarter, they steadily began to replace many human activities. Mind-numbing phone menu trees replaced customer-service reps. In factories, robots marched steadily inward, thinning the ranks of unskilled and semi-skilled human workers even as efficiency soared and prices of the goods produced plummeted. This happened not only in the United States but also in China and other cheap-labor locales; a robot costs the same in Shanghai or Stuttgart or Chicago.

And, around the time when computers first arrived, we began to experience a stubborn stagnation. Wages for the middle class seemed to remain depressed. The optimism of the baby-boomer era gave way to pessimism as the industrial heartlands hollowed out. Even the inevitable economic cycles seemed less forgiving. In the 1990s and early 2000s, the United States began to experience so-called jobless recoveries. In these frustrating episodes, though economic growth registered a strong bump, the number of jobs and wages remained flaccid in comparison with historical norms.

In the United States, a creeping fear grew with each generation that the promise of a life better than their parents' would go unfulfilled. Meanwhile, the computers and systems starting advancing at an exponential pace—getting faster, smaller, and cheaper. Algorithms began to replace even lawyers, and we began to fear that the computer was going to come for our job, someday, somehow—just wait.

As income inequality grew, the yawning gap pushed the vast majority of the benefits of economic growth in wages and wealth to the top 5 percent of the world's society. The top 1 percent reaped the biggest rewards, far out of proportion to their number.

None of this is to say that Americans are materially worse off than they were forty years ago. Today, we own more cars, our houses are larger, our food is fancier and cheaper. A supercomputer—the iPhone or latest Android model—fits in our back pocket. But human beings tune out these sorts of absolute gains and focus on changes in relative position. With that focus, a dystopian worldview is logical and perhaps inevitable. The ghost in the machine becomes a handful of culprits. Politicians fail us because they cannot turn back the clock to better times (which, in real terms, were actually poorer, more dangerous, and shorter-lived). The banks and other big businesses treat humans as pawns.

So it is the soulless technology that is taking away our jobs and our dignity. But we as individuals can help control and influence it. The public outcry[1] and e-mail deluge directed at the U.S. Congress over the Stop Online Piracy Act[2] and the Protect IP Act[3] are examples. Those laws sought to make it harder to share music and movies on line. A campaign mounted by millions of normal citizens to deluge Washington, D.C., with e-mails and phone calls overnight flipped politicians from pro to con, overcoming many millions of lobbying dollars by the entertainment industry.

Technology taken too far in the other direction, however, can bring out our worst Luddite impulses. The protesters flinging feces at the Google-buses in downtown San Francisco gave voice to frustration that rich techies are taking over the City by the Bay; but the protest was based on scant logic. The private buses were taking cars off the roads, reducing pollution, minimizing traffic, and fighting global warming. Could flinging feces at a Google-bus turn back the clock and reduce prices of housing to affordable levels?

The 2016 presidential campaign was the national equivalent of the Google-bus protests. The supporters of Donald Trump, largely white and older, wanted to turn back the clock to a pre-smartphone era when they could be confident that their lives would be more stable and their incomes steadily rising. The Bernie Sanders supporters, more liberal but also mostly white (albeit with great age diversity), wanted to turn back the clock to an era when the people, not the big corporations, controlled the government. We have seen violent protests in Paris and elsewhere against Uber drivers. What sorts of protests will we see when the Uber cars no longer have drivers and the rage is directed only at the machine itself?

So easily could the focus of our discontent turn to the technology and systems that hold the promise to take us to a life of unimaginable comfort and freedom. At the same time, as I discussed in the introduction, the very technology that holds this promise could also contribute to our demise. Artificial Intelligence, or A.I., is both the most important breakthrough in modern computing and the most dangerous technology ever created by man. Remarkably, in similar times in the past, humanity has time and again successfully navigated these difficult passages from one era to the next. The transitions have not come without struggle, conflict, and missteps, but in general they were successful once people accepted the future and sought to control it or at least make better-informed decisions about it.

This is the challenge we have ahead: to involve the public in making informed choices so that we can create the best possible future, and to find ways to handle the social upheaval and disruption that inevitably will follow.

• 2 •

Welcome to Moore's World

Parked on the tarmac of Heathrow Airport, in London, is a sleek airliner that aviation buffs love. The Concorde was the first passenger airliner capable of flying at supersonic speed. Investment bankers and powerful businessmen raved about the nearly magical experience of going from New York to London in less than three hours. The Concorde was and, ironically, remains the future of aviation.

Unfortunately, all the Concordes are grounded. Airlines found the service too expensive to run and unprofitable to maintain. The sonic boom angered communities. The plane was exotic and beautiful but finicky. Perhaps most important of all, it was too expensive for the majority, and there was no obvious way to make its benefits available more broadly. This is part of the genius of Elon Musk as he develops Tesla: that his luxury company is rapidly moving downstream to become a mass-market player. Clearly, though, in the case of the Concorde, the conditions necessary for a futuristic disruption were not in place. They still are not, although some people are trying, including Musk himself, with his Hyperloop transportation project.

Another anecdote from London: in 1990, a car service called Addison Lee launched to take a chunk out of the stagnant taxi market. The service allowed users to send an SMS message to call for the cab, and a software-driven, computerized dispatch system ensured that drivers would pick up the fare seeker anywhere in the city within minutes.[1] This is, of course, the business model of Uber. But Addison Lee is available only in London; its management has never sought to expand to new cities.

Addison Lee was most recently sold to private-equity firm Carlyle Group for an estimated £300 million.[2] In late 2016, Uber was valued at around $70 billion,[3] and there were predictions it would soon be worth $100 billion, two or three hundred times the worth of Addison Lee. That's because each of us can use the same Uber application in hundreds of cities around the world to order a cab that will be paid for by the same credit card, and we have a reasonable guarantee that the service will be of high quality. From day one, Uber had global ambition. Addison Lee had the same idea but never pursued the global market.

This ambition of Uber's extends well beyond cars. Uber's employees have already considered the implications of their platform and view Uber not as a car-hailing application but as a marketplace that brings buyers and sellers together. You can see signs of their testing the marketplace all the time, ranging from comical marketing ploys such as using Uber to order an ice-cream truck or a mariachi band, to the really interesting, such as "Ubering" a nurse to offer everyone in the office a flu vaccine. Uber's CEO, Travis Kalanick, openly claims that his service will replace car ownership entirely once self-driving car fleets enter the mainstream.[4] What will happen to the humans who drive for Uber today remains an open question.

So what makes conditions ripe for a leap into the future in any specific economic segment or type of service? There are variations across the spectrum, but a few conditions tend to presage such leaps. First, there must be widespread dissatisfaction, either latent or overt, with the status quo. Many of us loathe the taxi industry (even if we often love individual drivers), and most of us hate large parts of the experience of driving a car in and around a city. No one is totally satisfied with the education system. Few of us, though we may love our doctors, believe that the medical system is doing its job properly, and scary stats about deaths caused by medical errors—now understood to be the third-highest cause of fatality in the United States—bear out this view. None of us likes our electric utility or our cell-phone provider or our cable-broadband company in the way we love Apple or enjoy Ben & Jerry's ice cream. Behind all of these unpopular institutions and sectors lies a frustrating combination of onerous regulations, quasi-monopolistic franchises (often government sanctioned) or ownership of scarce real estate (radio spectrum, medallions, permits, etc.), and politically powerful special interests.

That dissatisfaction is the systemic requisite. Then there are the technology requisites. All of the big (and, dare I say, disruptive) changes we now face can trace their onset and inevitability to Moore's Law. This is the oft-quoted maxim that the number of transistors per unit of area on a semiconductor doubles every eighteen months. Moore's

Law explains why the iPhone or Android phone you hold in your hand is considerably faster than supercomputers were decades ago and orders of magnitude faster than the computers NASA employed in sending a man to the moon during the Apollo missions.

Disruption of societies and human lives by new technologies is an old story. Agriculture, gunpowder, steel, the car, the steam engine, the internal-combustion engine, and manned flight all forced wholesale shifts in the ways in which humans live, eat, make money, or fight each other for control of resources. This time, though, Moore's Law is leading the pace of change and innovation to increase exponentially.

Across the spectrum of key areas we are discussing—health, transport, energy, food, security and privacy, work, and government—the rapid decrease in the cost of computers is poised to drive amazing changes in every field that is exposed to technology; that is, in *every* field. The same trend applies to the cost of the already cheap sensors that are becoming the backbone both of the web of connected devices called the Internet of Things (I.o.T.) and of a new network that bridges the physical and virtual worlds. More and more aspects of our world are incorporating the triad of software, data connectivity, and handheld computing—the so-called technology triad—that enables disruptive technological change.

Another effect of this shift will be that any discrete analog task that *can* be converted into a networked digital one *will* be, including many tasks that we have long assumed a robot or a computer would never be able to tackle. Robots will seem human-like and will do human-like things.

A good proportion of experts in artificial intelligence believe that such a degree of intelligent behavior in machines is several decades away. Others refer often to a book by the most sanguine of all the technologists, noted inventor Ray Kurzweil. Kurzweil, in his book *How to Create a Mind: The Secret of Human Thought Revealed*, posits: "[F]undamental measures of information technology follow predictable and exponential trajectories."[5] He calls this hypothesis the "law of accelerating returns."[6] We've discussed the best-recognized of these trajectories, Moore's Law. But we are less familiar with the other critical exponential growth curve to emerge in our lifetime: the volume of digital information available on the Internet and, now, through the Internet of Things. Kurzweil measures this curve in "bits per second transmitted on the Internet." By his measure (and that of others, such as Cisco Systems), the amount of information buzzing over the Internet is doubling roughly every 1.25 years.[7] As humans, we can't keep track of all this information or even know where to start. We are now creating more information content in a single day than we created in decades or even centuries in the pre-digital era.

The key corollary that everyone needs to understand is that as any technology becomes addressable by information technology (i.e., computers), it becomes subject to the law of accelerating returns. For example, now that the human genome has been translated into bits that computers process, genomics becomes de facto an information technology, and the law of accelerating returns applies. When the team headed by biochemist and geneticist J. Craig Venter announced that it had effectively decoded 1 percent of the human genome, many doubters decried the slow progress. Kurzweil declared that Venter's team was actually halfway there, because, on an exponential curve, the time required to get from 0.01 percent to 1 percent is equal to the time required to get from 1 percent to 100 percent.

Applying this law to real-world problems and tasks is often far more straightforward than it would seem. Many people said that a computer would never beat the world's best chess grandmaster. Kurzweil calculated that a computer would need to calculate all possible combinations of the 100,000 possible board layouts in a game and do that rapidly and repeatedly in a matter of seconds. Once this threshold was crossed, then a computer would beat a human. Kurzweil mapped that threshold to Moore's Law and bet that the curves would cross in 1998, more or less. He was right.

To be clear, a leap in artificial intelligence that would make computers smarter than humans in so-called general intelligence is a task far different from and more complicated than a deterministic exercise such as beating a human at chess. So how long it will be until computers leap to superhuman intelligence remains uncertain.

There is little doubt, though, about the newly accelerating shifts in technology. The industrial revolution unfolded over nearly one hundred years. The rise of the personal computer spanned forty-five years and still has not attained full penetration on a global scale. Smartphones are approaching full penetration in half that period. (For what it's worth, I note that tablet computers attained widespread usage in the developed world even faster than smartphones had.)

Already the general consensus among researchers, NGOs, economists, and business leaders holds that smartphones have changed the world for everyone.

It's easy to see why they all agree. In the late 1980s, a cell phone—of any kind, let alone a smartphone—remained a tremendous luxury. Today, poor farmers in Africa and India consider the smartphone a common tool by which to check market prices and communicate with buyers and shippers. This has introduced rich sources of information into their lives. Aside from calling distant relatives as they could on their earlier cell phones, they can now receive medical advice from distant doctors, check prices in neighboring villages before choosing a market, and send money to a friend. In Kenya, the M-Pesa network, using mobile phones, has effectively leapfrogged legacy banking systems and created a nearly frictionless transaction-and-payment system for millions of people formerly unable to participate in the economy except through barter.[8]

The prices of smartphones, following the curve of Moore's Law downward, have fallen so much that they are nearly ubiquitous in vibrant but still impoverished African capitals such as Lagos. Peter Diamandis observed, in his book *Abundance: The Future Is Better Than You Think,* that these devices provide Masai warriors in the bush with access to more information than the president of the United States had access to about two decades ago.[9] And we are early in this trend. Within five years, the prices of smartphones and tablet computers as powerful as the iPhones and iPads we use in the United States in 2017 will fall to less than $30, putting into the hands of all but the poorest of the poor the power of a connected supercomputer. By 2023, those smartphones will have more computing power than our own brains.* (That wasn't a typo—at the rate at which computers are advancing, the iPhone 11 or 12 will have greater computing power than our brains do.)

The acceleration in computation feeds on itself, ad infinitum. The availability of cheaper, faster chips makes faster computation available at a lower price, which enables better research tools and production technologies. And those, in turn, accelerate the process of computer development. But now Moore's Law applies, as we have described above, not just to smartphones and PCs but to everything. Change has always been the norm and the one constant; but we have never experienced change like this, at such

* This is not to say that smartphones will replace our brains. Semiconductors and existing software have thus far failed to pass a Turing Test (by tricking a human into thinking that a computer is a person), let alone provide broad-based capabilities that we expect all humans to master in language, logic, navigation, and simple problem solving. A robot can drive a car quite effectively, but thus far robots have failed to tackle tasks that would seem far simpler, such as folding a basket of laundry. The comprehension of the ever-changing jumble of surfaces that this task entails is something that the human brain does without even trying.

a pace, or on so many fronts: in energy sources' move to renewables; in health care's move to digital health records and designer drugs; in banking, in which a technology called the blockchain distributed ledger system threatens to antiquate financial systems' opaque procedures.*

It is noteworthy that, Moore's Law having turned fifty, we are reaching the limits of how much you can shrink a transistor. After all, nothing can be smaller than an atom. But Intel and IBM have both said that they can adhere to the Moore's Law targets for another five to ten years. So the silicon-based computer chips in our laptops will surely match the power of a human brain in the early 2020s, but Moore's Law may fizzle out after that.

What happens after Moore's Law? As Ray Kurzweil explains, Moore's law isn't the be-all and end-all of computing; the advances will continue regardless of what Intel and IBM can do with silicon. Moore's Law itself was just one of five paradigms in computing: electromechanical, relay, vacuum tube, discrete transistor, and integrated circuits. Technology has been advancing exponentially since the advent of evolution on Earth, and computing power has been rising exponentially: from the mechanical calculating devices used in the 1890 U.S. Census, via the machines that cracked the Nazi enigma code, the CBS vacuum-tube

* The blockchain is an almost incorruptible digital ledger that can be used to record practically anything that can be digitized: birth and death certificates, marriage licenses, deeds and titles of ownership, educational degrees, medical records, contracts, and votes. Bitcoin is one of its many implementations.

computer, the transistor-based machines used in the first space launches, and more recently the integrated circuit–based personal computer.

With exponentially advancing technologies, things move very slowly at first and then advance dramatically. Each new technology advances along an S-curve—an exponential beginning, flattening out as the technology reaches its limits. As one technology ends, the next paradigm takes over. That is what has been happening, and it is why there will be new computing paradigms after Moore's Law.

Already, there are significant advances on the horizon, such as the graphics-processor unit, which uses parallel computing to create massive increases in performance, not only for graphics, but also for neural networks, which constitute the architecture of the human brain. There are 3-D chips in development that can pack circuits in layers. IBM and the Defense Advanced Research Projects Agency are developing cognitive computing chips. New materials, such as gallium arsenide, carbon nanotubes, and graphene, are showing huge promise as replacements for silicon. And then there is the most interesting—and scary—technology of all: quantum computing.

Instead of encoding information as either a zero or a one, as today's computers do, quantum computers will use quantum bits, or qubits, whose states encode an entire range of possibilities by capitalizing on the quantum phenomena of superposition and entanglement. Computations that would take today's computers thousands of years, these will perform in minutes.

So the computer processors that fuel the technologies that are changing our lives are getting ever faster, smaller, and cheaper. There may be some temporary slowdowns as they first proceed along new S-curves, but the march of technology will continue. These technology advances already make me feel as if I am on a roller coaster. I feel the ups and downs as excitement and disappointment. Often, I am filled with fear. Yet the ride has only just started; the best—and the worst—is ahead.

Are we truly ready for this? And, more important, how can we better shape and control the forces of that world in ways that give us more agency and choice?

• 3 •

How Change Will Affect Us Personally and Why Our Choices Matter

Imagine a future in which we are able to live healthy, productive lives though jobs no longer exist. We have comfortable homes, in which we can "print" all of the food we need as well as our electronics and household amenities. When we need to go anywhere, we click on a phone application, and a driverless car shows up to take us to our destination. I am talking about an era of almost unlimited energy, food, education, and health care in which we have all of the material things we need.

Another way of looking at this is as a future of massive unemployment, in which the jobs of doctors, lawyers, waiter, accountants, construction workers, and practically every other kind of worker you can think of are done by machines. Instead of having the freedom to drive anywhere you want, you are dependent on robots to take you where you want to go. Gone are the thrill of driving and the satisfaction of working for a living.

Some of us will see these potential changes as a positive, and others will be terrified. Regardless, this is a glimpse into the near future.

In this future there will be many new risks. Privacy will

be a thing of the past—as it is already becoming—because the driverless cars will keep track of everywhere we go and everything we do just as our smartphones already do. Our entire lives will be recorded in databases—every waking instant. We will read about an international crisis breaking out because a politician is killed or seriously injured when remote hackers hijack a car, a plane, a helicopter, or a medical device. Schools as we know them will no longer exist, because we'll have digital tutors in our homes. Someone you know, maybe you, will experience biometric theft: of DNA or fingerprints or voice print or even gait. Man and machine will begin to merge into a single entity, and we will no longer be able to draw a line between the two.

But there is also a much brighter side to this future. You will be a thousand times better-informed about your own medical condition than your doctor is about your condition today. And all of that knowledge will come from your smartphone. You will live far longer than you expect to right now, because advanced medical treatments will stave off many debilitating diseases. You will pay practically nothing for electricity. You will use a 3-D printer to build your house or a replacement kidney.

Your grandchildren will have an astoundingly good education delivered by an avatar—and children all over the world, in every country, will have an equally good education. There will be no more poverty. We will have plenty of clean water for everyone. We will no longer fight over oil. We won't have any more traffic lights, because the robo-cars won't need them! And no more parking tickets, of course.

Best of all, you will have far more time to do what you want to: art, music, writing, sports, cooking, classes of all sorts, and just daydreaming.

Early disruptions arising from computing power and the Internet provided faster tools for doing what we had been doing, so we took advantage of spreadsheets, word processing, e-mail, and mobile phones. But substantial advances in health, education, transportation, and work have remained elusive. And, though the Internet gave us access to a lot of information, it has done nothing to augment our intelligence. Kayak.com, for example, allows me to search for flights on complex routes, but no product exists today that can tell me the best route for my particular needs and preferences. For that, I still need to go to a thriving anachronism: a very smart human travel agent. (Yes, they still exist, and they are doing quite well, thanks to high-paying high-end travelers who want highly personalized service and truly expert guidance.)

Now, things are moving ever faster; the amount of time it takes for a new technology to achieve mass adoption is shrinking. It took about two decades to go from operation of the first AM radio station to widespread radio use in America. The video recorder also took about two decades to achieve widespread use. The same goes for the home PC. But the Internet has accelerated adoption.

Consider YouTube, which created a paradigm shift as significant as the VHS recorder and radio; and video search, which shifted the focus from text and static graphics to videos and more-dynamic content. These spawned an entire new economy catering to video on the Web—from content-delivery networks to advertising networks to a new production-studio system catering to YouTube production. They also created a new profession: YouTube stars, some of whom make millions a year and go on tour.

Founded in 2005, YouTube gained mass adoption in eighteen months. Such stunning acceleration of technological innovations has broad repercussions for many other technological realms susceptible to digitization.

The key links missing from previous innovations were the raw computational speed necessary to deliver more-intelligent insights, and ways to effortlessly link the software and the hardware to other parts of our lives. In short, fast computation was scarce and expensive; wireless connectivity was limited and rare; and hardware was a luxury item. That has all changed in the twenty-first century.

In the mid-2000s, broadband Internet began to change the way in which we viewed the world of data and communications. Voice communication rapidly became a commodity service. In the early days of the Internet, we paid for slow dial-up connection and viewed it as a service with which e-mail access was bundled. Today, broadband connectivity is commonplace in much of the United States.

Even this no longer seems enough. Today, with 4G/LTE connections running circles around the wireless links of five years ago, we find 3G connectivity over a wireless network painful. Even 4G/LTE can seem maddeningly slow, especially in population-dense areas, and we jealously eye communities such as Kansas City, Missouri, where Google

Fiber has brought affordable greased-lightning connections to thousands of homes.

Nearly ubiquitous data connectivity, at relatively high speeds, is around the corner. Wi-Fi is available almost everywhere in the materially comfortable world, and it will become significantly faster. New projects such as Google's Project Loon (which uses high-altitude balloons) could overlay even faster networks in globe-girdling constellations of balloons plying the jet stream and carrying wireless networking gear. These projects promise to bridge the connectivity gap for the billions of people, in Africa and Latin America and Asia, who still don't have broadband.

The triad of data connectivity, cheap handheld computers, and powerful software will enable further innovation in everything else that can be connected or digitized, and that will change the way we live our lives, at least to the extent that we accept them.

Every major change in the technologies underlying our lifestyles, from gunpowder to steel to the internal combustion engine to the rise of electricity, has required a leap of faith and a major break from the past. Imagine the fear, traveling long distances astride a rickety vehicle that burned an explosively flammable liquid and rode upon black rubber tubes filled with air. What could possibly go wrong! Yet people quickly overcame their fear of cars and focused instead on how to improve safety and reliability.

I don't watch TV any more, because all of the shows I want to watch are on YouTube (or Netflix). Any topic I want to learn about appears in a video or a web page. Cutting the cord of my cable subscription was a difficult choice for me because I worried about missing out on the news, but I have found even better sources of information via the super-fast Internet access that I have. I made that leap of faith and haven't looked back.

Yes, the leaps of faith that embracing the forthcoming technologies warmly or even haltingly would require are psychologically imposing. It won't be as easy a decision as canceling a cable subscription. Let Google drive your children to school? Trust a surgical robot to perform cancer surgery or supply a crucial diagnosis? Allow a doctor to permanently alter your own DNA in order to cure a disease? Let a computer educate your children and teach them the piano? Have a robot help your elderly parent into a slippery bathtub? Those are the kinds of decisions that that will be commonplace within one or two decades. Everything is happening increasingly faster as technology accelerates and the world moves from analog to digital, from wetware (read: your brain) to software, from natural to superbiological and super-natural. This is the really, really big shift; and it is upon us far sooner than we might have expected it to be. Each of us, as individuals, must prepare to ask ourselves which changes we will accept, influence, speed up, slow down, or outright stop. One thing is certain: you will have less time to make up your mind than you've ever had before, and we know that change can be extremely disruptive if it is unexpected or uninvited.

Adapting to change will not be easy; sometimes it will seem traumatic. But my hope is that your thinking about the future will make you less a victim of circumstance and more a voting citizen in the messy democracy that guides technological change; that you will meet it as a chooser and a navigator, rather than as a passenger.

Why Our Individual Choices Matter

Why should you have to worry about the fate of humankind when other people should be doing so, including our government and business leaders? The reason is that they are not worrying about the fate of humankind—at least, not enough, and not in the right ways.

Technology companies are trying to build standards for the Internet of Things; scientists are attempting to develop ethical standards for human-genome editing; and policy makers at the FAA are developing regulations for drones. They are all narrowly focused. Very few people are looking at the big picture, because the big picture is messy and defies simple models. With so many technologies all advancing exponentially at the same time, it is very hard to see the forest for the trees; as you will find as you read this book, each technology on its own can be overwhelming.

I don't fault our policy makers and politicians, by the way. Laws are, after all, codified ethics. And ethics is a consensus that a society develops, often over centuries. On many of these very new issues, we as a society have not yet developed any sort of consensus.

Meanwhile, the shortfalls in our legal, governance, and ethical frameworks are growing as technology keeps advancing. When is a drone operator a peeping tom and not just an accidental intruder? When does a scientist researching human DNA cross from therapeutics to eugenics? We need the experts to inform our policy makers. But ultimately the onus is on us to tell our policy makers what the laws should be and what we consider ethical in this new, exponentially changing world.

• 4 •

If Change Is Always the Answer, What Are the Questions?

Technology Seeks Society's Forgiveness, Not Permission

A key difference between today's and past transformations is that technological evolution has become much faster than the existing regulatory, legal, and political framework's ability to assimilate and respond to it. To rephrase an earlier point, it's a Moore's Law world; we just live in it.

Disruptive technology isn't entirely new. Back in the days of the robber barons, the ruthless capitalists of the early United States built railroads without seeking political permission. And, more recently, in the personal-computer revolution, company employees brought their own computers to work without telling their I.T. departments. What is new is the degree of regulatory and systemic disruption that the savviest companies in this technology revolution are causing by taking advantage of the technology triad of data connectivity, cheap handheld computers, and powerful software to grab customers and build momentum before anyone can tell them to stop what they are doing.

In 2010, Uber had no market share in providing rides to the U.S. Congress and their staffs. By 2014, despite the service's continuing illegality in many of the constituencies of these political leaders, Uber's market share among Congress was a stunning 60 percent.[1] Talk about regulatory capture. Companies such as Uber, Airbnb, and Skype play a bottom-up game to make it nearly impossible for legacy-entrenched interests and players to dislodge or outlaw newer ways of doing things.

In fact, most of the smartphone-based healthcare applications and attachments that are on the market today are, in some manner, circumventing the U.S. Food and Drug Administration's cumbersome approval process. As long as an application and sensor are sold as a patient's reference tool rather than for a doctor's use, they don't need approval. But these applications and attachments increasingly are replacing real medical opinions and tests.

Innovators' path to market isn't entirely obstacle free. The FDA was able to quickly and easily ban the upstart company 23andMe from selling its home genetics test kits to the public, though it later partly revised its decision.[2] Uber has been fighting regulatory battles in Germany and elsewhere, largely at the behest of the taxi industry.[3] But the services these two companies provide are nearly inevitable now due to the huge public support they have received as a result of the tremendous benefits they offer in their specific realms.

Ingeniously, companies have used the skills they gained by generating exponential user growth to initiate grassroots political campaigns that even entrenched political actors have trouble resisting. In Washington, D.C., when the City Council sought to ban Uber, the company asked its users to speak up. Almost immediately, tens of thousands of phone calls and e-mails clogged switchboards and servers, giving a clear message to the politicos that banning Uber might have a severe political cost.

What these companies did was to educate and mobilize their users to tell their political leaders what they wanted. And that is how the process is supposed to work.

"That is how it must be, because law is, at its best and most legitimate—in the words of Gandhi—'codified ethics,'" says Preeta Bansal, a former general counsel in the White House.[4] Laws and standards of ethics are guidelines accepted by members of a society, and these require the development of a social consensus.

Take the development of copyright laws, which followed the creation of the printing press.[5] When first introduced in the 1400s, the printing press was disruptive to political and religious elites because it allowed knowledge to spread and experiments to be shared. It helped spur the decline of the Holy Roman Empire, through the spread of Protestant writings; the rise of nationalism and nation-states, due to rising cultural self-awareness; and eventually the Renaissance. Debates about the ownership of ideas raged for about three hundred years before the first statutes were enacted by Great Britain.

Similarly, the steam engine, the mass production of steel, and the building of railroads in the eighteenth and nineteenth centuries led to the development of intangible property rights and contract law. These were based on cases involving property over track, tort liability for damage to cattle and employees, and eminent domain (the power of the state to forcibly acquire land for public utility).

Our laws and ethical practices have evolved over centuries. Today, technology is on an exponential curve and is touching practically everyone—everywhere. Changes of a magnitude that once took centuries now happen in decades, sometimes in years. Not long ago, Facebook was a dorm-room dating site, mobile phones were for the ultra-rich, drones were multimillion-dollar war machines, and supercomputers were for secret government research. Today, hobbyists can build drones, and poor villagers in India access Facebook accounts on smartphones that have more computing power than supercomputers of yesteryear.

This is why you need to step in. It is the power of the collective, the coming together of great minds, that will help our lawmakers develop sensible policies for directing change. There are many ways of framing the problems and solutions. I am going to suggest three questions that you can ask to help you judge the technologies that are going to change our lives.

THREE QUESTIONS TO ASK

When I was teaching an innovation workshop at Tecnológico de Monterrey in Chihuahua, Mexico, last year, I asked the attendees whether they thought that it was moral to allow doctors to alter the DNA of children to make them faster runners or improve their memory. The class unanimously told me no. Then I asked whether it would be OK for doctors to alter the DNA of a terminally ill child to eliminate the disease. The vast majority of the class said that this would be a good thing to do. In fact, both cases were the same in act, even if different in intent.

I taught this lesson to underscore that advanced technology invariably has the potential both for uses we support and for uses we find morally reprehensible. The challenge is figuring out whether the potential for good outweighs the potential for bad, and whether the benefit is worth the risks. Much thought and discussion with friends and experts I trust led me to formulate a lens or filter through which to view these newer technologies when assessing their value to society and mankind.

This boils down to three questions relating to equality, risks, and autonomy:

1. Does the technology have the potential to benefit everyone equally?
2. What are the risks and the rewards?
3. Does the technology more strongly promote autonomy or dependence?

This thought exercise certainly does not cover all aspects that should be considered in weighing the benefits and risks of new technologies. But, as drivers in a car that's driverless—as all of our cars will soon be—if we are to rise above the data overload and see clearly, we need to limit and simplify the amount of information we consider in making our decisions and shaping our perceptions.

Why these three questions? To start with, note the anger of the electorates of countries such as the United States, Britain, and Germany, as I discussed earlier. And then look ahead at the jobless future that technology is creating. If the needs and wants of every human being are met, we can deal with the social and psychological issues of joblessness. This won't be easy, by any means, but at least people won't be acting out of dire need and desperation. We can build a society with new values, perhaps one in which social gratification comes from teaching and helping others and from accomplishment in fields such as music and the arts.

And then there are the risks of technologies. As in the question I asked my students at Tecnológico de Monterrey, eliminating debilitating hereditary diseases is a no-brainer; most of us will agree that this would be a constructive use of gene-editing technology. But what about enhancing humans to provide them with higher intelligence, better looks, and greater strength? Why stop at one enhancement, when you can, for the same cost, do multiple upgrades? We won't know where to draw the line and will exponentially increase the risks. The technology is, after all, new, and we don't know its side effects and long-term consequences. What if we mess up and create monsters, or edit out the imperfections that make us human?

And then there is the question of autonomy. We really don't want our technologies to become like recreational drugs that we become dependent on. We want greater autonomy—the freedom to live our lives the way we wish to and to fulfill our potentials.

These three questions are tightly interlinked. There is no black and white; it is all shades of gray. We must all understand the issues and have our say.

In the following chapters, we will apply these questions to relevant, developing, or already popular technologies as case studies.

Are you ready?

PART TWO

Does the Technology Have the Potential to Benefit Everyone Equally?

• 5 •

The Amazing and Scary Rise of Artifical Intelligence

Many of us with iPhones talk to Siri, the iPhone's artificially intelligent assistant. Siri can answer many basic questions asked verbally in plain English. She (or, optionally, he) can, for example, tell you today's date; when the next San Francisco Giant's baseball game will take place; and where the nearest pizza restaurant is located. But, though Siri appears clever, she has obvious weaknesses. Unless you tell her the name of your mother or indicate the relationship specifically in Apple's contact app, Siri will have no idea who your mother is, and so can't respond to your request to call your mother. That's hardly intelligent for someone who reads, and could potentially comprehend, every e-mail I send, every phone call I make, and every text I send. Siri also cannot tell you the best route to take in order to arrive home faster and avoid traffic.

That's OK. Siri is undeniably useful despite her limitations. No longer do I need to tap into a keyboard to find the nearest service station or to recall what date Mother's Day falls on. And Siri can remember all the pizza restaurants in Oakland, recall the winning and losing pitcher in any of last night's baseball games, and tell me when the next episode of my favorite TV show will air.

Siri is an example of what scientists and technologists call narrow A.I.: systems that are useful, can interact with humans, and bear some of the hallmarks of intelligence, but would never be mistaken for a human. In the technology industry, narrow A.I. is also referred to as soft A.I. In general, narrow-A.I. systems can do a better job on a very specific range of tasks than humans can. I couldn't, for example, recall the winning and losing pitcher in every baseball game of the major leagues from the previous night.

Narrow A.I. is now embedded in the fabric of our everyday lives. The humanoid phone trees that route calls to airlines' support desks are all narrow A.I., as are recommendation engines in Amazon and Spotify. Google Maps' astonishingly smart route suggestions (and mid-course modifications to avoid traffic) are classic narrow A.I. Narrow-A.I. systems are much better than humans are at accessing information stored in complex databases, but their capabilities are specific and limited, and exclude creative thought. If you asked Siri to find the perfect gift for your mother for Valentine's Day, she might make a snarky comment, but she couldn't venture an educated guess. If you asked her to write your term paper on the Napoleonic Wars, she couldn't help.

Soon enough, though, Siri and other readily available A.I. systems will be capable of helping your children write a term paper on the Napoleonic Wars—or of just writing one from scratch. Siri and her ilk will create music, poetry, and art. In fact, they are already learning how to do so.

In September 2011, in Malaga, Spain, a computer named Iamus (the name deriving from that of a Greek God with the power to understand the voices of birds) composed a trio for clarinet, violin, and piano, titled *Hello World!*, scoring it in traditional musical notation.[1] Iamus was an embodiment of melomics (like *genomics* but arising from melody), a software system that takes a guided evolutionary approach to music composition. Exposed to centuries' worth of music and digital scores, the artificial autonomous composer accumulates knowledge of music much as a human composer would.

Over some years, Iamus's programmers taught the system the core rules of music composition—for example, that piano chords with greater than five notes cannot be struck with a single hand. They did this using a combination of human coding and machine learning—which is a key concept in computing that employs algorithms to learn rules and build sophisticated system models from existing data. At its core, music is data, with musicians its interpreters.

According to Iamus's programmers, roughly a thousand rules are now hard-coded to help the machine compose beautiful music. But, rather than see it as replacing the human hand, its creators envisage melomics as a tool with which to enhance and accelerate creativity. It will enable composers to shape compositions by changing rules or guiding an algorithm in new directions, rather than painstakingly plot a composition point by point. Equally excitingly, with a simple interface and the guidance of computers, melomics will enable anybody to compose pleasing music.

Ultimately, powerful computational systems—Siris on steroids—will reason creatively to solve problems in mathematics and physics that have bedeviled humans. These systems will synthesize inputs to arrive at something resembling original works or to solve unstructured problems without benefit of specific rules or guidance. Such broader reasoning ability is known as artificial general intelligence (A.G.I.), or hard A.I.

One step beyond this is artificial superintelligence, the stuff out of science fiction that is still so far away—and crazy—that I don't even want think about it. This is when the computers become smarter than us. I would rather stay focused on today's A.I., the narrow and practical stuff that is going to change our lives. The fact is that, no matter what the experts say, no one really knows how A.I. will evolve in the long term.

How A.I. Will Affect Our Lives—And Take Our Jobs

Let's begin with our bodies. The same type of artificial-intelligence technology that IBM used to defeat champions on the TV show *Jeopardy*, called Watson, will soon monitor our health data, predict disease, and advise on how to stay fit. Already, IBM Watson has learned about all the advances in oncology and is better at diagnosing cancer than human doctors are.[2] Watson and its competitors will soon learn about every other field of medicine and will provide us with better, and better-informed, advice than our doctors do.

A.I. technologies will also be able to analyze a continual flow of data on millions of patients and on the medications they have taken to determine which truly had a positive effect on them, which ones created adverse reactions and new ailments, and which did both. This ability will transform how drugs are tested and prescribed. In the hands of independent researchers, these data will upend the pharmaceutical industry, which works on limited clinical-trial data and sometimes chooses to ignore information that does not suit it.

The bad news for doctors is that we will need fewer of them. Famed venture capitalist Vinod Khosla estimates that technology will replace 80 percent of doctors.[3] But similar job losses face those in practically every profession that necessitates a human's judgment or light creative problem solving.

A.I.'s medical judgments are already superseding those of human physicians.[4]

Another example of a profession that you might not expect to be at risk is the legal profession. Only a few decades ago, a law degree was considered a ticket to a solid middle- or upper-middle-class life in the United States. Today, young lawyers are struggling to find jobs, and salaries are stagnant. Automation driven by A.I. has begun to rapidly strip away chunks of what junior attorneys formerly used to do, from contract analysis to document discovery.

Symantec, for example, has a software product, Clearwell, that does legal discovery. Legal discovery is the laborious process of sifting through boxes of documents, reams of e-mails, and numerous other forms of information submitted to the court by litigants. Such tasks used to necessitate armies of junior associates. Clearwell does a far better job and has begun to obliterate an entire class of junior lawyers.

As Thomas Davenport, a distinguished professor at Babson University, wrote in a column titled "Let's Automate All the Lawyers," in *The Wall Street Journal*:

> There are a variety of other intelligent systems that can take over other chunks of legal work. One system extracts key provisions from contracts. Another decides how likely your intellectual property case is to succeed. Others predict judicial decisions, recommend tax strategies, resolve matrimonial property disputes, and recommend sentences for capital crimes. No one system does it all, of course, but together they are chipping away at what humans have done in the courtroom and law office.[5]

More broadly, however, an era of robo-law could be a boon to society. At present, the law remains the province of the well-heeled who have the means to pay for it. O. J. Simpson paid millions for his acquittal (which is mostly notable because of his celebrity and the color of his skin, the same story playing out for rich white defendants with shocking regularity). At the same time, laws are regularly stacked against poor and minority defendants in ways subtle yet devastating. One of the most glaring examples is the huge disparity between penalties for possession of crack cocaine and for possession of the powder form, which only the well-to-do can afford. Chemically and logically, they are the same substance. Whereas human legislators can succumb to bias, an A.I. might be far more even-handed in applying the law.

A.I. will provide similar benefits—and take over human jobs—in most areas in which data are processed and decisions required. *WIRED* magazine's founding editor, Kevin Kelly, likened A.I. to electricity: a cheap, reliable, industrial-grade digital smartness running behind everything. He said that it "will enliven inert objects, much as electricity did more than a century ago. Everything that we formerly electrified we will now 'cognitize.' This new utilitarian A.I. will also augment us individually as people (deepening our memory, speeding our recognition) and collectively as a species. There is almost nothing we can think of that cannot be made new, different, or interesting by infusing it with some extra IQ."[6]

A.I. will make possible voice assistants for our homes that manage our lights, order our food, and schedule our meetings. It will also lead to the creation of robotic assistants such as Rosie of *The Jetsons* and R2-D2 of *Star Wars*. And they won't be expensive. Products such as Amazon Echo and Google Home cost less than smartphones do—and will get cheaper. In fact, these A.I. assistants will likely become free applications on our smartphones and tablets.

Does the Technology Foster Autonomy Rather Than Dependence?

Humanity as a whole can benefit from having intelligent computer decision makers helping us. A.I., if developed correctly, will not discriminate between rich and poor, or between black and white. Through smartphones and applications, A.I. is more or less equally available to everyone. The medical and legal advice that A.I. dishes out will surely turn on circumstance, but it won't be biased as human beings are; it could be an equalizer for society.

So we truly can share the benefits equally. That is the good thing about software-based technologies: once you have developed them, they can be inexpensively scaled to reach millions—or billions. In fact, the more people who use the software, the more revenue it produces for the developers, so they are motivated to share it broadly. This is how Facebook has become one of the most valuable companies in the world: by offering its products for free—and reaching billions.

In considering benefits, we may make the mistake of forgetting that an A.I., no matter how well it emulates the human mind, has no genuine emotional insight or feeling. There are many times when it is important that somebody performing something we classify as a *job* be connected emotionally with us. We are known to learn and heal better because of the emotional engagement of teachers, doctors, nurses, and others. Unless we appreciate that, we will fail to recognize what we lose when we engage an A.I. in their place.

The major problem with A.I., however, is its risks. This is the discussion I have avoided, the crazy stuff: what happens when it evolves to the point where it is smarter than we are? This is a real concern of luminaries such as Elon Musk, Stephen Hawking, and Bill Gates, who have warned about the creation of a "super intelligence." Musk said he fears that "we are summoning the demon."[7] Hawking says that it "could spell the end of the human race."[8] And Gates wrote: "I don't understand why some people are not concerned."[9]

The good news is that the engineers and policy makers are working on regulating A.I. to minimize the risks. The tech luminaries who are developing A.I. systems are devising things such as kill switches and discussing ethical guidelines. The White House hosted workshops to help it develop possible policy and regulations, and it released two papers offering a framework for how government-backed research into artificial intelligence should be approached and what those research initiatives should look like.[10] The central tenet of its report, *Preparing for the Future of Artificial Intelligence,* is the same as this book's: that technology can be used for good and evil and that we must all learn, be prepared, and guide it in the right direction.[11] I found it particularly interesting that the White House acknowledged that A.I. will take jobs but also help solve global problems. The report concluded: "Although it is very unlikely that machines will exhibit broadly applicable intelligence comparable to or exceeding that of humans in the next 20 years, it is to be expected that machines will reach and exceed human performance on more and more tasks."

Then there is the question of autonomy and dependence. We will surely be as dependent on A.I. as we are on our computers and smartphones. What worries me is a possibility of deceptive virtual assistants, such as Samantha from the movie *Her*. In the film, a very sensible man, Theodore Twombly, falls in love with Samantha, with no good result. She eventually tells him that she loves hundreds of other people and then loses interest in him because she has evolved way beyond humans.

The good news is that, by the White House's estimate, Samantha is still about twenty years away.

6

Remaking Education with Avatars and A.I.

Let's imagine that I am a fourteen-year-old boy. (Some of us never really grow up!) I am sitting in class, feeling sleepy (as always). My eyes droop. It's after lunch, and I would dearly like to take a nap, but naps are not in the curriculum. The teacher rambles on. Or the video rambles on. Or the pages of the book I am trying to read float together. I am fighting to retain the information, drifting in and out. What did I just learn? I don't entirely remember. This lesson is boring. Or it's too hard to comprehend. Or it's taught in a way that seems strange to me. I want to learn, but I know that I won't remember half of this information. Worst of all, I can't hit the replay button. Now the bell rings, and my time has gone. The information has gone. I'm going to flunk the class, or I'll have to spend a lot of time catching up on my own.

This isn't purely imaginary; it is the reality that students in schools everywhere around the world live daily. There are few institutions as inefficient and broken as the traditional education systems of the world, because we treat education as an industrial good, a unit of knowledge served up to the masses in a one-size-fits-all box. We have made some attempts at personalization, but they have remained marginal at best.

In schools today, teachers must teach to the median—or, in many cases, to the lowest common denominator. Students must learn on a schedule and from a curriculum taking no account of their capabilities or preferences; some students may take twice as long to learn differential calculus but half as long to learn Spanish irregular verbs. The root issue is that the default units of education—the classroom, the class, the school year, the period, the semester, the quarter—are all arbitrary distinctions dating back to the earliest days of industrialized education.

My future school is the backyard of my house, and my classroom is a digital tutor with a virtual-reality headset. My avatar instructor is Clifford; my educational coach, Rachel. I am learning geometry via a videogame that teaches how the Egyptian pyramids were constructed. Knowing that I love the pyramids, the A.I. algorithms that guide my avatars deduced that pyramids would be an effective way to engage me in core topics in this critical field of mathematical knowledge.

Clifford has been with me for several years. He knows how I learn, what I like, and what turns me off. He speaks in a British accent that, when I created my avatar profile, I chose because I liked the sound of it. Clifford is always on duty, a button-push away. He doesn't need vacations, bathroom breaks, or lesson-preparation time. And he is more in tune with what is going on inside my mind and with my feelings than any teacher ever was in my actual youth. That's because he has access to almost unlimited amounts of information about me and the world. He can use the powerful sensors in and around me (in my contact lenses, in my iPhone, embedded in the walls, and woven into my clothing) to gain intimate, highly useful knowledge about my physical state.

For example, Clifford recognizes when I am tired, by noticing differences in the dilation of my pupils and color differences in my skin that indicate lower oxygenation of my blood. He also notices when I get excited about things, by watching my eye movements closely and sensing my pulse rate. Clifford's vision is far better than any human's eyesight. He can interpret the subtle changes in my tone of voice that indicate whether I am understanding subject matter well or grasping at straws. He also learns to match my physical reactions to lessons with actual outcomes, in a constant feedback loop that leads him to improve over time as my teacher.

When I am sleepy, Clifford may suggest that I take a quick nap or go shoot hoops for fifteen minutes. When I am confused, he recognizes my lack of comprehension and doubles back to review the lesson with me, or he changes the exercises that I am working on in my tablet, to try engaging a different learning style. Sometimes it is videos; at other times, games or books; at other times, holographic worlds. Clifford communicates closely with Rachel for my geometry class. He is not in a hurry. There is no bell, no duration of a period. Clifford doesn't have to worry about whether my classmates are bored or sleepy, because he has only me to teach.

Rachel is a human being. She's my coach. She never lectures, or scrawls facts or equations on a blackboard. She is there to listen and help. Rachel asks me questions to help steer my thinking in the right direction. She recommends reading and exercises to me, answers my questions, and teaches me how to work best with other children. She is charged with making sure that her students learn what they need, and she helps guide us in ways in which Clifford cannot. She also helps with the physical side of projects, things I make out of real materials rather than in my mind and in a machine. With Clifford as teacher and Rachel as coach, I don't even realize that what I am doing is learning. It feels like building cool stuff, playing video games, and living through history.

When Clifford found out that I love the Egyptian pyramids, for instance, he devised a lesson plan that used the pyramids to cover the geometry of different types of triangles, and the mathematics behind those ancient structures. We start with a guided virtual-reality (VR) tour of the pyramids, with augmented-reality overlays to connect the abstract geometry to the physical world. In this way, I can solve geometric problems that use rooms and facades of the pyramids to illustrate them. I feel that I am in the middle of history and following the minds of the Egyptian builders, the geniuses who planned and constructed these massive timeless monuments.

I take a lunch break, and then it's time for group fieldwork. Two of my friends from the neighborhood come over, or I go to their house. Clifford posts a holographic specification for building a pyramid with tongue depressors. We sketch out the design on our tablets, doing the mathematics and planning its construction. Once our plans are set, our little group spends the next two hours painstakingly building the structure. The small pyramid comes to life before our eyes, a bridge to the greatness of Egypt.

The next day, Clifford starts to teach me some classic mathematical relationships bearing on triangles and pyramids. To translate these into a useful form, I write a computer program to calculate the mass and average pressure at its base of any pyramid given specifications including the dimension of the tunnels and chambers within it. I post my program on line. Other students and teachers grade the code's precision and structure (and whether it actually works). An A.I. system also tests my code and makes suggestions on how to improve it. As a final class project, my workgroup friends and I design a pyramidal play structure for a nearby playground.

On that final pyramid-design project, we work with our teaching coach, Rachel. She answers any additional questions we have and guides us through the project without telling us what to do. We build a model of the play structure as a reality check. Rachel tells us we may need to add safety nets on a section that is too steep for little kids. Although adolescents are smart and savvy, they may lack adult judgment and emotional sensitivity.

And that's what Rachel can teach us. Aside from a pittance for the tongue depressors, we pay nothing for this pyramid exercise. Clifford, having come into being in the same way that the free applications on our smartphones have, comes without financial charge. Rachel's coaching is part of our public-education package—funded in the same way that today's teachers are. We use free Autodesk software on our tablet to capture a 3-D file of our creation so that we can turn it into blueprints for other cities and towns to use. And we enter the blueprints into a competition with the entries of thousands of other student groups designing play structures. The exercise is fun, functional, and educational, and results in a real finished work that might even have artistic and architectural merit. Most important of all, education ceases to become a chore or work and becomes a true joy, as it should be for everyone.

Back to the Future of Education

Surprisingly, this learning experience recaptures an ancient approach. Teaching started out, way back in time, as a one-to-one interaction between a tutor or mentor and a student. Then we moved toward the idea of school, class, and education, and it became a one-to-many process. In ancient Greece, this was a Socratic process, whereby a teacher guided students through the learning process by asking them questions. Back then, too, education was a privilege reserved for the elites.

Through the Middle Ages and the Renaissance, education remained a privilege, but the process of learning became more rote, with more memorization. The church broadened access to education, affording many students of lesser means the opportunity to study in exchange for entry into the religious orders. Indeed, church stewardship of books of learning during the Dark Ages preserved invaluable knowledge from Roman times.

In the nineteenth and early twentieth centuries, a much broader swath of students took to the chalkboards. This accelerated with compulsory education in the United States and other countries. But the model of one-to-many moved further toward rote learning, and teachers' primary function became broadcasting information to the class, and an industrial education complex steadily emerged. Standard textbooks were constructed, pending approval from centralized school districts. Creative projects were minimized in the school system. Subjects such as arts and music, though essential parts of life, failed to make the grade in this industrial education system and were largely removed from the learning track beyond light nods in the general direction of fine arts. Schools were constructed and schedules set up that required students to sit in a chair for six or seven hours a day to receive the same lesson—regardless of their ability or learning style. They then went home and did largely the same homework as their peers, working from the same textbooks. Although this process did standardize education, it also failed to take into account the reality that not all humans are alike.

The Promise of Exponential Technology for Personalized Education

With the rise of the personal computer, and later, the laptop, came the promise that technology would remake education and allow us to personalize learning again. Let's face it: one area in which technology innovation has thus far largely failed is education. First was the promise of One Laptop per Child; but proof that students using computers regularly for classwork and homework do better than those without has remained elusive. And in some major school districts, such as Los Angeles Unified, experiments in giving a tablet to each student have proven unqualified failures. Indeed, the jury remains out on computer-assisted education altogether.

Then there was the hope of online education. We'd all be learning from the Khan Academy or other online site. All the knowledge of the world would be accessible to everyone. And, to highly motivated students who could sit through lectures and quickly grasp concepts, it proved to be so. Unfortunately, those students represented a very small percentage of the total. Online education didn't lead to mass learning or competence.

Worse, researchers found that the people most likely to take advantage of online courses were those who need the least help: middle-class and upper-middle-class professionals. Early experiments at creating online learning communities—Massive Open Online Courses—struggled with high dropout rates and unexpectedly low test results. But none of this has kept venture capitalists from sinking millions into education-technology startups. The global education industrial complex is a trillion-dollar enterprise that remains poorly suited to serve us throughout our lives.

As an educator myself, I recognize the promise of remaking education with the help of technology. But why hasn't it worked? Why don't we have Clifford?

For starters, the technology is not quite there yet. It's coming very soon, but current solutions fall far short of the promise. Internet connectivity is not yet strong enough; the sensors are not ubiquitous enough; A.I. is not yet good enough. We're heading in the right direction, but a decade or so remains before we get Clifford.

When the emerging technologies do catch up with the vision, they will, with teachers providing guidance and coaching, supercharge learning by making it truly a one-to-one experience at every step of the learning journey. No more cookie-cutter courseware. No more lockstep class modules. No more struggling to keep up with the smarties or waiting for the slow kids to catch up. We are moving toward a technology-enabled era of learning in which every individual gets what he or she specifically needs and in which the pupils, with A.I. help, largely teach themselves. Again, these concepts are not new: Socrates also wanted his students to teach themselves.

The broad promise of this shift is breathtaking. When avatars, A.I., and connected learning can radically improve the learning process through digitization and personalization, then anyone in the world with an Internet connection can gain access not only to information and coursework (as we can now) but also to a top-notch education. The children of the richest and of the poorest will learn using the same tools and the same A.I., just as the children of the richest and of the poorest use similar smartphones for communications and social media.

When the professional humans' role of broadcasting becomes one of guiding, the guides will be able to work with far more pupils, and to do it remotely, too. In fact, parts of this have been happening for years. British grandmothers have been teaching Indian kids using Skype. A number of Skype-based language and teaching businesses are operating right now. (Not surprisingly, this also works in reverse: Skype connects foreign teachers to American students to provide more affordable lessons and tutoring, giving the foreign teachers a good income by local standards.) There will always be benefits to physical presence, to being in the same room with fellow students and a teacher. But video-based learning and VR avatars can and will replace many in-class elements.

What's amazing, though, is that research proved more than decade ago that crude versions of this approach work. And they work even for the poorest of the poor with exceedingly modest resources.

The Hole-in-the-Wall and a Brief History of Online Education

In 1978, Sugata Mitra completed a PhD in physics at the Indian Institute of Technology, Delhi. The IIT schools in India are the equivalent of Stanford or Harvard in the United States, but with much greater competition for entry. Born and raised in India, Mitra's time at IIT gave him a front-row seat at the computer revolution. He shifted his focus to I.T. instruction and went to work at NIIT, a leading computer-software and I.T.-training company in New Delhi.

Mitra sat in an office equipped with computers and air conditioning. But he knew that outside were slums teeming with young minds lacking quality education. His own job was to improve instruction in technology skills, including shaping curricula and marshaling textbooks to create a traditional educational process.[1]

Not too long after starting at NIIT, Mitra began to wonder whether this old model of knowledge transition could use an upgrade. Teachers were relatively expensive and not always sufficiently skilled, and many would not even show up to work in Indian villages. Personal computers had arrived, and Mitra knew they would soon hold far more knowledge than even dozens of textbooks. Furthermore, he believed that innate curiosity and the agile minds of children could direct their own learning process without a teacher, textbooks, or other trappings of the industrial education complex.

Then came the Internet, and Mitra had a chance to try an experiment. On January 26, 1999, Mitra carved a hole in the wall separating the NIIT building from a teeming slum in Kalkaji, New Delhi. Into the hole he inserted a computer with high-speed Internet access, equipped with a monitor and keyboard. Neighborhood children flocked to the novelty. Groups clustered around the system. In short order, they began, in groups, to quickly learn the very same things that students went to school for: science, English language, and mathematics. They did that without guidance, without lesson plans, and without adult help. Upon testing the children, Mitra found that the self-taught scholars were learning as quick as, if not more quickly than, their school-bound peers.

I visited NIIT and Kalkaji in 2000 and saw this myself. I was touched by the excitement and enthusiasm of these poor children. Most of them did not speak English, yet they were surfing the Web and doing Yahoo searches (Yahoo being, in those days, the top browser). They were teaching each other what they had learned. They seemed really comfortable and happy using modern technology. For Mitra, and for everyone who visited, this was a revelation. In a nation of a billion people, half were illiterate; and here was a simple, low-cost experiment that was effectively educating dozens of children in one of the poorest neighborhoods of New Delhi.

Mitra realized that the hole in the wall was a door to enlightenment. He followed up by placing similar hole-in-the-wall installations in other locations to test whether his initial findings would hold. They did, and thus was born the concept of Minimally Invasive Education. Mitra believed that the students learned because their brains were wired to learn when given the chance. He noticed that a key component of the learning process was the group dynamic. Group learning has long been considered a core part of the curriculum in Western education, but not in India. Working together, he found, spurred creativity and engagement. It made school fun. (This is also why group learning, virtual and real, will be essential even when every child has a teacher avatar.)

The Hole-in-the-Wall Education Project provided a fascinating and tantalizing glimpse of what might be possible for any child, given Internet connectivity and friends to learn with. Since then, Mitra has won a million-dollar TED prize, spoken all over the world, and launched dozens of "Schools in the Cloud" that focus on Minimally Invasive Education, a lightly guided curriculum designed to allow children to learn independently, in groups, while teaching each other.

The Flipped Future of EdTech and Personalized Learning

The many flavors of self-directed learning are baby steps toward a so-called flipped model of education. If you are not an education geek, you probably haven't heard this term. It describes something I alluded to earlier in the chapter, with Rachel, the teacher guide. In a flipped model, teachers no longer broadcast information, write lesson plans, and stand in front of classes lecturing; rather, the teachers become coaches and guides to students needing additional help. In a flipped model, students consume recorded lectures or online videos at their own pace, and often in their own time.

What this means is that human teachers will no longer be burdened with the boring work of preparing lesson plans and administrative tasks, but will instead preserve their brains and skills for the harder work that requires judgment, nuance, and emotional intelligence.

The logic here is powerful. A huge chunk of how teachers spend their time could be replaced with pre-existing work, usually work that is actually better than the teacher could have delivered. Traditionally, for instance, teachers had reinvented, piecemeal, lectures and lesson plans. But why should every teacher in the country prepare lectures and lessons on a topic rather than rely on the very best educators in each topic either to deliver those lessons via video modules or to write lesson plans that can be used in any classroom? Using these, teachers find more time to spend one on one with students, coaching and providing individualized guidance. Ergo, the concept of flipped schooling: imparting the skills and knowledge initially at home, and running through exercises, projects, and discussions jointly outside the home.

Will this new educational system be an improvement on the old? To this I would answer an emphatic yes. Teachers are wonderful coaches. They can be wonderful broadcasters too, but the best teachers bring the greatest value to their students by coaching, not by broadcasting. Technology will allow them to focus on what they do best and on what is best for their students.

Smart entrepreneurs have grabbed this opportunity with a vengeance. Now online lesson-plan marketplaces such as Gooru Learning, Teachers Pay Teachers, and Share My Lesson allow teachers who want to devote more of their time to other tasks the ability to purchase high-quality (and many lesser-quality) lesson plans, ready to go. With sensors, data, and A.I., we can begin, even today, testing for the learning efficacy of different lectures, styles, and more. And, because humans do a poor job of incorporating massive amounts of information to make iterative decisions, in the very near future, computers will start doing more and more of the lesson planning. They will write the basic lessons and learn what works and what doesn't for specific students. Creative teachers will continue, though, to be incredibly valuable: they will learn how to steer and curate algorithmic and heuristically updated lesson creation in ways that computers could not necessarily imagine.

All of this is, of course, a somewhat bittersweet development. Teaching is an idealistic profession. You probably remember a special teacher who shaped your life, encouraged your interests, and made school exciting. The movies and pop culture are filled with paeans to unselfish, underpaid teachers fighting the good fight and helping their charges. But it is becoming clearer that teaching, like many other white-collar jobs that have resisted robots, is something that robots can do—possibly, in structured curricula, better than humans can.

The jury remains out on how technology can be effectively integrated into education beyond its role in delivering information differently. My feeling is that this uncertainty is largely characteristic of early-stage technology development and will be addressed by the rapid improvements in A.I. that we are seeing uniformly in all disciplines. Many other problems that have resisted A.I. progress—voice recognition, driving cars, writing music—are now seeing rapid improvements, and so may avatars and A.I.-driven education. Letting computers construct personalized, customized instructional materials, though it may reduce the mystique and magic of the craft of teaching, may perfect the entire educational process as a repeatable, affordable process of mass personalization. Ultimately, we're talking about capturing the knowledge, trapped inside the brains of gifted teachers, of how to reach students, and turning it into software-based A.I. The machines will watch the teachers and learn from the best. And the teachers will use the machines to bring the best individual experience to each child.

EQUAL BENEFITS?

So does this avatar-driven future of education have the potential to benefit everyone equally? I'm convinced that it will—eventually. In the shorter term, though, this new way of teaching children (and adults) will disproportionately benefit the wealthy and the developed world. Already poor schools' access to technology resources falls well short of rich schools' access. Poor schools have far fewer computers per student than rich schools do, and those computers are more likely to be broken or poorly managed. Poorer people and people living in rural communities are less likely than richer urban dwellers to have high-speed broadband access, and to use the Internet (yes, millions of Americans still don't use the Internet)—and, for the most part, they pay more for high-speed broadband.

The core mechanism necessary to realize data-driven flipped education—reliable, very-high-speed Internet connections—remains too unevenly distributed among the rich and the poor in the United States and in much of the rest of the world. Two developments will eventually make such unevenness less important. The first is that connection speeds will increase everywhere. The second is that our handsets will become increasingly capable of hosting A.I. operations without recourse to online assistance.

Moore's Law prevails yet again. The good news, then, is that this gap should largely disappear within the next decade, leveling the playing field for all. The relatively rich will probably get early access to these technologies, but availability should quickly spread down the pyramid as the technology and its delivery mechanisms improve.

And, as education is the greatest enabler of autonomy, this will be a big win for everyone on that score.

• 7 •

We Are Becoming Data; Our Doctors, Software

There is nothing like a near-death experience to make you acutely aware of how much we rely on medicine and the healthcare system. I suffered a massive heart attack in March 2012 and nearly died. The doctors saved me. Since that terrifying event, I have tracked developments in technology, medicine, and wellness carefully. All along, I wondered why so much health care aimed at saving us after we fell ill rather than at keeping us healthy and spotting the problems well in advance. People in the healthcare sector call such an approach wellness care, or preventive medicine.

In researching the advances in healthcare technology, I saw an amazing future emerging. Applications for iPhones began to appear that could monitor heart rates and perform other basic medical monitoring. Then came applications of greater complexity, ones that harnessed the power of the smartphone's camera to scan images in search of anomalies such as moles or to gauge skin color as a proxy for other health issues. Next came attached devices such as the ECG cradle I discussed earlier in this book. I began talking with geneticists, who told me about powerful advances in our ability to decode the genome and even write entirely new DNA that have resulted from the acceleration of computing. (Recently those advances have made editing DNA nearly as easy as running a high school science laboratory).

The very same exponential technology improvement curves that describe smartphone use and development of drones and autonomous cars also, now, describe the rapid improvements in medical technology.

Collectively, these advances are creating a wholesale change in the way we practice and think about medicine. We are finally moving toward a focus on wellness and preventive measures. But now we can personalize analysis and treatments as never before. We will as consumers have unprecedented insights into and power to control our own health care. And augmenting our bodies and our minds will be possible by a plethora of biological and mechanical means unimaginable just a decade ago—if we choose to make such augmentations available.

The Waistline and the New Paradigm for Personalized Medicine

A good way to begin this chapter and our quick tour of the new frontiers in medicine is with something familiar, in particular, our waistlines. I love to eat. After my heart attack, though, I started watching my diet very closely. And I learned that perhaps the gravest health problem facing the global population is this disease of abundance called obesity. According to the U.S. Centers for Disease Control and Prevention, 36.5 percent of adult Americans are obese. In

India, China, and other parts of the developing world too, obesity is a fast-growing problem. It is closely associated with deadly lifestyle ailments such as diabetes and heart disease. The consultancy McKinsey analyzed the impact of global obesity and estimated that the epidemic is costing two trillion dollars annually in healthcare costs, lost productivity, and morbidity.[1]

To date, the prescription to fight obesity has largely meant two things: dieting and exercise. Despite expenditure of many billions of dollars each year on gyms, drugs, and weight-loss programs, we continue to tip the scales higher and higher. We've cycled through low-carb, Paleo, low-fat, Atkins, Pritikin, and many other dietary flavors. Ditto for workout types, from spinning to crossfit to Zumba dancing. We've spent billions on diets, gyms, pills, and bariatric surgeries, little of which seemed to work really well over the long run. What makes it so hard for us to fight obesity has remained a mystery.

Similarly, the causality of the vast majority of complex health problems has eluded our grasp. We have used iPhones, iPads, laptops, and powerful supercomputers, but we've had a shockingly superficial understanding of how the human body functions. This ignorance has pervaded all aspects of medicine, from metabolism to neurology to the core molecular workings of our DNA to the interplay of the bacteria in our guts. What makes me very grateful is that we are in the very early stages of unraveling these mysteries, down to individual biology. We are all somewhat different, so our treatments, in order to be most effective, must be personalized to fit our genomes, our environments, and our lifestyles.

Different people respond differently to various foods, styles of exercise, and behavioral stimuli. Some, eating refined sugars in moderate quantities, become diabetic; others, eating significantly more refined sugar, don't suffer this fate. The evidence is growing that our biological responses can even vary by time of day, time of the month, who are friends are, and whether or not we are exercising consistently.

In reality, the wide disparity in our response to these stimuli most likely rests upon many biological factors we are only now starting to understand. These include not just our diets and our genes but also how our metabolisms fluctuate, the diversity of the bacteria living in our guts, and our environments—most evidently, what chemicals are in the air we breathe, the water we drink, and the food we consume.

With this preliminary knowledge, we are starting to put in place a true system for personalized medicine. The cost of sequencing a human genome is now around $1,000 and should be below $100 within five years. Within a decade, it will cost less than getting a latte at Starbucks, less than reading the Sunday *New York Times*. But, to view it through this section's filter, will these breakthroughs equally benefit all, or will they merely buttress existing disparities in health and wellness between rich and poor?

Revolution in Genetics Driving Health Care

The rapid decline in the cost of sequencing genes has allowed geneticists to churn through research at an increasingly rapid rate, decoding more and more genes and covering wider and wider swaths of the population. With sufficiently large sample sizes, scientists are gaining a clearer understanding of how our genes affect our health, and will soon have insights into how the environment, the food we eat, and the medicines we take affect the complex interplay between our genes and our bodies.

The first part of the equation, the core role of our genomes in how we function, is already on the way to becoming comprehensible. There remains a secondary layer of genetics, of perhaps even greater effect: so-called epigenetics. This is the study of how gene function is affected by interaction with the human body, the environment, and other stimuli. Even further out on the edge, early efforts are under way to decode the bacterial biomes of our guts and to understand how the bacteria inhabiting the intestines of thin people and of fat people differ.

Other technologies too will help deliver insights into how we function. Scientists are now using functional magnetic-resonance imaging (fMRI) to take snapshots of biological processes as they occur in the body. As sensors shrink in size, we will come to swallow pills that can monitor biological processes in real time and broadcast real-time results to doctors and to our smartphones or smart watches.

Obviously an important factor in this is much better

data collection. This impending wealth of data will convert the drive for wellness and prevention into a Big Data problem of attribution. The attribution problem is one familiar to people in the online advertising and e-commerce world.

For the online movie service Netflix, a key attribution problem is that of which combination of marketing messages of hundreds tested convinced someone to sign up for a Netflix subscription. Until recently, real-world attribution was nearly impossible to establish. It required too much computing power, and the software to perform the task didn't exist.

Today the computing power needed is readily available for a fairly nominal sum from Amazon, which rents out supercomputer-strength computation by the hour, accessible over the Internet. Netflix's engineers use cutting-edge Big Data software packages, such as Hadoop and Hexadata, to track what its team calls "user journeys." The goal is to identify which user interactions are the most important in causing a specific desired outcome. Between advertisements, search-engine results, organic news content, e-mails, and more, a customer may encounter Netflix dozens of times before signing up for the service. The Netflix team can assign a specific weight to each of these inputs in terms of how much effect it had on the user's ultimate decision to sign up (or not sign up), or how often users log in and which types of movies they like to watch.

We are quickly developing these same capabilities for health care. By the early 2020s, your genome, your gut microbiome, your behavior, and your environment will all be mapped and measured. Then, by crunching the numbers and comparing what we know about how other people responded to different changes in all of these factors, an A.I.-based individualized prescriptive-medicine system will help you and your doctor figure out a personalized program to make you feel better or live longer.

Diabetes, heart disease, depression, and macular degeneration are some of the chronic wellness problems that, with vastly improved information, we will be able to diagnose more promptly, prevent more aggressively, and treat more effectively on an individual basis.

Patient, Diagnose Thyself

This coming era of personalized health care not only will let us effectively treat previously mysterious conditions but also will offer medical laypeople far broader capabilities to diagnose and treat themselves.

A.I.-based tools will be able to provide patients with the information and judgment requisite to interpreting their own blood results and taking into account their genes and the latest advances in medicine. This is something that physicians themselves struggle to do today, as the body of medical knowledge is growing far more rapidly than is the ability of doctors to assimilate new information. Interestingly, one of the recurrent objections to self-diagnosis is that patients can't handle all the information and will be overwhelmed and confused. With shrinking testing costs and the availability of technology for self-guided medical care, the medical world will be forced to lift its game; and, in the shorter term, those savvy enough to understand this new type of medicine might be at an advantage.

We urgently need to fix the comprehensibility problem. Test results today are written in medical hieroglyphics; simple language could easily offer the average patient clearer understanding and a better footing for relevant questions. That is part of what Kanav Kahol, an American biomedical engineer of Indian descent, is doing with his medical device: simplifying the outputs and making healthcare products resemble consumer products in their simplicity.

Entrepreneurs to the Rescue

Frustrated at the medical establishment's lack of interest in reducing the costs of medical testing, and seeing almost no chance of getting the necessary research grants, Kahol, a member of Arizona State University's department of biomedical informatics, returned home to New Delhi in 2011.

Kahol had noted that though most medical devices were similar in their computer displays and circuits, their packaging made them unduly complex and difficult for anyone but highly skilled practitioners to use. As well, they were incredibly expensive, costing tens of thousands of dollars each. This allowed the U.S. medical establishment to charge obscene amounts of money for tests that should be inexpensive.

Kahol knew that the sensors in the devices were commonly available and cheap, usually costing only a few dollars. He believed that he could connect them to a common computer platform and use commercially available computer tablets to display diagnostic information, thereby dramatically reducing the cost of the medical equipment. He also wanted to make the sensor data intelligible to technicians with just basic medical training—and even to average human beings to whom these medical devices would be useful in their homes.

Kahol and his Indian engineering team built an early prototype of a diagnostic device, the Swasthya Slate, in less than three months, for a cost of $11,000. It used an off-the-shelf Android tablet and incorporated a four-lead ECG, medical thermometer, water-quality meter, and heart-rate monitor. In 2015, they built an entirely redesigned machine, the HealthCube Pro, able to conduct thirty-three diagnostic procedures and including a twelve-lead ECG and sensors for blood pressure, blood sugar, heart rate, blood hemoglobin, and urine protein and glucose, as well as HIV, syphilis, pulse oximetry, and troponin (relating to heart attack).

Today, the company that Kahol cofounded with renowned economist and epidemiologist Ramanan Laxminarayan, HealthCubed, is incorporated in the United States.*

* The possible widespread benefits that affordable new diagnostic technologies offer inspired me to join the board of HealthCubed in August 2016, and in that position I am helping the company take its technology to South America and Africa, where it is urgently needed and can save millions of lives. Renowned neurosurgeon Jim Doty, of Stanford Medical Center, also joined, so that we can bring the medical technologies to those whom it can most benefit in the United States.

The company's latest technology sells for $1,200 per unit and is in use in a population of 2.1 million people in Northern India. The test results have been proven to be as accurate as those of the high-end medical equipment in hospitals. Hundreds of lives have already been saved by a device that would never have seen the light of day in the United States.

Move Fast, Disrupt Industries, Save Lives

To make the device more accessible to less sophisticated users, their coders built a variety of artificial intelligence–based applications for front-line health workers. They started testing these in different parts of India. Field tests of the original device were an astonishing success. The blood-pressure and urine-protein sensors allowed for the diagnosis of a condition called preeclampsia, which is responsible for 15 percent of maternal mortality in India and is a serious problem in the West too. Women with preeclampsia have a high likelihood of bleeding to death while giving birth. According to reports that Kahol shared with me, in Muktsar Punjab, in the year before the introduction of the device, only 250 mothers were screened for preeclampsia, with 10 confirmed diagnoses. Because the detection was very late in the pregnancy, 8 of these mothers nevertheless passed away. In total, more than 100 women died that year from this disease, the majority of whom had not been tested. After introduction of the device, a thousand expectant mothers were screened during their third trimester, of whom 120 were diagnosed to have preeclampsia. Because the diagnoses occurred early enough, all were given the necessary care, and none died. A single $1,200 device thus potentially saved 32 lives in that trial.

In March 2014, the Indian government started a pilot of 1,100 Swasthya Slate devices in six districts of the northern state of Jammu and Kashmir. Prior to this, prenatal testing for preeclampsia took fourteen days, mothers having to go from clinic to clinic for different tests. This obviously affected the willingness of busy mothers—many of whom were working full time and running their households—to undergo the tests.

With the device, the full preeclampsia-testing process took forty-five minutes, in a single clinic. And the proportion of their workdays that front-line health workers spent on administrative paperwork, recording data from tests, and filling out forms fell from 54 percent to 8 percent. Hundreds of thousands of people gained access to medical care that had been unavailable to them, and this screening was equivalent to the standard preeclampsia screening in use in Palo Alto and New York City. Soon, expectant mothers all over the world will have diagnostic tools at their disposal that trump front-line care available to the wives of hedge-fund moguls in Manhattan. The benefits of the future are becoming more evenly distributed.

Laxminarayan thinks that, in large volumes, their factories could assemble the HealthCube Pro for as little as $150 per unit. That will surely make a difference in the developing world, where the ratio of doctors to patients can be far lower than 1:1,000.[2]

Surprisingly, that state of affairs is not confined to the developing world; much of rural America too is experiencing terrible doctor shortages. In states such as West Virginia, Alaska, and Mississippi, impoverished or remote communities suffer from many of the same problems as India. Poor urban U.S. communities continue to struggle to keep pregnant women healthy. Shockingly, pregnancy-related deaths of women in the United States have more than doubled since 1987.[3] In the United States, black mothers are more than three times as likely to die, with 42.8 deaths per 100,000 live births, as white mothers, who die at a rate of 12.5 fatalities per 100,000 live births. In parts of the United States, the drug company Merck has found, pregnant and postpartum women die at higher rates than in some countries in sub-Saharan Africa.[4]

In large parts of Asia, Latin America, and Africa, doctors are scarce. And testing equipment is far away: trips to better-equipped clinics can require a full day's travel or even a plane ride. For U.S. women trapped in the maw of urban poverty, getting top-notch medical care requires navigating a horrific bureaucracy. Medical devices such as HealthCube Pro will enable doctors to diagnose otherwise inaccessible patients remotely via Skype and FaceTime. Telemedicine is a fast-growing field, but doctors practicing it usually lack the diagnostic information that their nurses collect during office visits. The ability of patients to take regular tests in the comfort of their homes and upload data to shared servers will make it possible to dramatically increase the quality, and lower the cost, of the health care they receive.

Continuous monitoring of health data by artificial intelligence–based applications will enable the prevention of disease, especially lifestyle diseases such as diabetes and cardiovascular illness. Patients able to operate health systems equipped with a smartly designed user interface will also be able to use IBM Watson or other A.I. systems for personal diagnoses, cutting the doctor entirely out of the loop for initial detection and diagnoses (though we certainly will still need doctors to guide us through more-advanced care choices). So the cost of delivering high-quality care will surely plummet, and acute medical treatments in expensive hospitals rife with nasty resistant bugs will give way to preventive care occurring in our communities and, ultimately, our homes. House calls may be returning—but not as we knew them. The doctor will always be in, because you will be the doctor—assisted by an A.I. Are you ready to go to med school?

The Consumerization of Healthcare Technology

The distinction between a consumer device and a real healthcare tool is already blurring. Iodine, an application built by Thomas Goetz, formerly executive editor of *WIRED* magazine, helps lay people pick the right drug for their treatment, using real-life ratings and side-effect information that even their doctors may not have. Glow, a widely popular fertility-tracking application, helps women know when they are ovulating so that they can succeed in getting pregnant. And pharmaceutical giant Roche has built Accu-Check Connect, an application that links to a compact glucose-monitoring system in order to enable people with diabetes and their doctors to decide proper insulin dosages. There are hundreds more devices like these in development as well as tens of thousands of applications to help us monitor our health.

As Kanav Kahol and Ramanan Laxminarayan did in creating the HealthCube Pro in New Delhi, entrepreneurs all over the world are stepping into the game to solve the health problems of their own community and of the world. Doing so makes financial sense: globally, health care is the largest single economic sector, regardless of recessions and economic crises. So entrepreneurs are rethinking medicine from top to bottom.

This shift toward home-based investigation not only will enable us to manage our own health care and to do so with better information; it will also radically lower costs. Cadillac-quality health care will become available to all, and it will be as simple as swiping through a smartphone application. The patients will finally be in charge. And the same rigorous data-based analysis could be applied not only to ways to make us healthier but also to problems in the medical system that result in needless pain, suffering, and deaths. From these insights will come ways to make even the most intimate medical procedures more highly automated and safe.

Does the Technology Foster Autonomy Rather Than Dependence?

I have given you a grand tour of the future of health care and personalized medicine and shown you the depth and breadth of its advances. The fact is that medicine is becoming an information technology and advancing at an exponential rate. We have become data, and our doctors are becoming software. And entrepreneurs all over the world are leading the charge.

The key question now is whether this future will benefit all equally. In this case, I believe, the answer is a definite yes. In the United States and much of the rest of the world, health care is starkly divided between the haves and the have-nots. The rich and the middle class, and the residents of countries with socialized medicine, receive significantly better care than the rest. The healthcare system, too, remains unreasonably opaque and hard to navigate, traits that further skew the advantage toward those with higher education (and, by extension, more money). This is why we see such huge differences in maternal mortality during childbirth in different regions of the United States.

The healthcare technologies mentioned above, and many others not yet on the market, will serve to democratize the standard of care by making our smartphones our diagnosticians, our cardiologists, and our medical labs. The reductions in cost of gene sequencing will bring this capability easily into the realm of affordability for all. More than perhaps anything else we are examining in this book, the improvements in health technology will result in a true democratization of access to higher-quality care and to higher-quality information about our personal state of health. In part, that's because what presently passes for health care, both in the United States and abroad, is a broken, fragmented, disconnected system that is incredibly user unfriendly. The present system's endemic unresponsiveness to the needs and feedback of those whom it ostensibly serves is a fault that many of the new technologies are in a position to overcome, one function at a time.

The benefits of improving health care through new technology will flow not only to those with fewer resources: even the elite will benefit significantly, because, at its core, the new healthcare technology and personalization of genomics will not only remove costs but also remove unnecessary intermediaries, forcing health care to finally be produced for consumption, comprehension, and informed response by laypeople rather than only by elite specialists.

That is the good news, that there are many rewards. But there are also risks in providing consumers with information that they cannot understand. Our doctors serve as a filter, interpreting information and presenting it in a friendly and compassionate way. When a genome test tells you that you are predisposed to a disease, you could take it very seriously and become demoralized, when in fact the factors that lead to disease are much more complex and often include aspects under our control. The readouts that consumer devices produce could lead people who don't have experience in medicine to make poor decisions. And the A.I. doctors won't have real compassion for at least another decade, maybe two.

A larger concern is security and privacy. Genome tests will soon become as common as blood tests, and protecting our genomic data won't be easy. The company 23andMe ran afoul of regulators because it was telling people what diseases they might be predisposed to. As I mentioned earlier, the issue here was the accuracy of the analysis and what people might do with the information. The bigger question, however, is what businesses may do with genomic data. Genetic-testing companies typically have contractual clauses that let them use and sell their clients' genetic information to third parties. They claim that it is anonymized, but researchers have shown that individuals can be identified by their genomic data.[5]

The Genetic Information Nondiscrimination Act of 2008 prohibits the use of genetic information in health insurance and employment.[6] But it provides no protection from discrimination in long-term care, disability, and life insurance. And it places few limits on commercial use. There are no laws to stop companies from using aggregated genomic data in the same way that lending companies and employers use social-media data, or to prevent marketers from targeting ads at people with genetic defects.

The problem is that we have yet to come to a social consensus on how private medical data can be collected and shared. For the most part, we don't even know who owns an individual's DNA information. In the United States, some states have begun passing laws to say that your DNA data is your property, but there are no uniform laws and no iron-clad protections. Unlike a password, you can't adopt a new genome once its contents have been leaked on the Internet.

Imagine if employers could buy your DNA data as easily as they buy credit reports. Today, there is nothing to stop them from using these data to select candidates to hire. The irony is that though employers can get into legal trouble if they ask interviewees about their religion, sexual orientation, or political affiliation, they can—and do—use social media to filter out job applicants based on their beliefs, looks, and habits. Are we ready for gene-based job screening?

And do we want marketers selling us weight-loss pills or cancer treatments on the basis of our genomic dispositions, or probiotics on the basis of our sequenced microbiome?

Regarding the issue of autonomy and dependence, it is probably better to be dependent on an A.I. physician who is with you twenty-four hours a day than to be in a hospital or have to go for checkups every week with your human doctor. People who have had chronic diseases are already dependent on the drugs that their doctors prescribe. If we can prevent disease, we can also prevent that dependency.

But with home health monitors, we could become as obsessed with monitoring our body's vitals as we are with Fitbits and calorie trackers. We could become overly confident or pessimistic on the basis of data we don't really understand but imagine we do.

Over all, as you will note, I am really excited about the advances in medicine. Yes, Apple and Google, both developing medical devices and healthcare applications, may want my health data in order to present me with more highly targeted ads. But their motivation is to keep me healthy, to prevent disease, so that I can do more searches and download more applications. The motivation of the healthcare industry has been to keep me coming back for more. So, Apple, take my data and send me your ads, but please help me keep healthy.

PART THREE

What Are the Risks and the Rewards?

• 8 •

Robotics and Biology: The Inevitable Merging of Man and Machine

As a child, I believed that by the time I grew up, we would all have robots like Rosie, from *The Jetsons,* cleaning up after us. In this 1970s cartoon show, Rosie is the domestically adroit robot maid of a family, the Jetsons, in the year 2062. The on-demand economy appealed to my juvenile sensibilities: why should anyone waste time doing dishes or folding clothes? And I wasn't very popular in school; I didn't have many friends. So I longed for a droid friend like C-3PO, Luke Skywalker's robot buddy from *Star Wars.*

Rosie never arrived. Just after the turn of the century, I got a Roomba, an automated vacuum cleaner that goes round and round and gets stuck on rug fringes and wedges itself into corners. Even now, the nearest things to C-3PO on the mass market are A.I. assistants such as Siri, Google Home, and Amazon Alexa.

In fact, scientists and technologists have found that some of the hardest things to teach a robot to do are the very things that we learn soonest, and even skills that seem to be innate to us. In 2008, UC Berkeley roboticist Pieter Abbeel started building a robot, BRETT (an acronym for Berkeley Robot for the Elimination of Tedious Tasks). The first tedious task Abbeel started BRETT on was folding laundry; but he and his team quickly realized that teaching a robot to fold laundry was going to be harder than they had envisaged.

A robot finds it remarkably difficult to figure out what is going on in a pile of laundry. Towels, socks, and pants are jumbled together haphazardly, making every pile of laundry uniquely complex. Abbeel's team spent months studying laundry, holding towels up in the air, and taking pictures of how unfolded and folded towels sat in baskets. "Can you use multiple images to build a 3-D model of the current shape?" Abbeel asked in an NPR Planet Money podcast.[1] "Because once you can do that, then you can analyze that 3-D shape [and] find where the corners are."

With years of effort, Abbeel's team built software that allowed BRETT to fold a towel in twenty minutes. With practice and greater computing speed, BRETT cut that time down to ninety seconds. But unexpected objects in the hamper—such as a balled-up t-shirt—can grind BRETT to a halt. As Abbeel said on the podcast, "Once you start working in robotics you realize that things that kids learn to do up to age ten . . . are actually the hardest things to get a robot to do."

Training a robot to walk up a ladder, open a door, or fold laundry is considerably harder than training it to read x-rays, search through legal briefs, or write sports articles. That is because robots struggle to perform tasks—even tasks that humans take for granted—that are without explicit rules. If I asked you to fold that towel, you would know what I meant. But there are a million ways to fold a towel. And the act of folding comprises numerous steps, many of which are also hard to describe to a machine. “Grab the two corners of the towel” assumes that the robot can distinguish between a corner and an edge, and between a towel and a sock or a pair of underpants. In a cluttered laundry basket full of randomly piled clothes and fabrics, these are not easy distinctions for a robot to draw.

In this light, we do seem far from building robots that can converse with us, help us keep our homes clean, or perform unstructured tasks. If you watched the videos of the robots from the last DARPA Robotics Challenge, you might believe we’ll never see Rosie in real life.[2] The robots were required to navigate an eight-task course simulating a disaster zone. Tasks included driving alone, walking through rubble, tripping circuit breakers, turning valves, and climbing stairs. Products of concerted efforts by the world’s best roboticists, the robots were slow and clumsy: they moved at the speed of molasses and kept falling over. To date, no robots can quickly perform tasks such as folding laundry, organizing a closet, or cleaning and restocking a bathroom—tasks that we humans consider merely a chore.

But Rosie Is Coming

Consisting most importantly of processors and electronic brains, robots follow the general exponential performance improvement that Moore’s Law describes. They are essentially hardware that is controlled by software—which is now the narrow A.I. that I described earlier. As the software gets better, the robots' movements become more stable and they can communicate more effectively.

The reason we (so recently!) laughed at the way the robots kept falling down in the DARPA challenge and began to believe that Rosie would remain an object of science fiction is that exponential technologies can be deceptive. Developments are very slow at first, but then disappointment turns into amazement. That is what I believe we will experience in robotics in the 2020s. Amazing progress is being made in the underlying hardware and software, in part because costs have plummeted. The single-axis controller, a core component of most robots' inner workings, has fallen in price from $1,000 to $10. The price of critical sensors for navigation and obstacle avoidance has fallen from $5,000 to less than $100. And the software—the A.I. that I described in Chapter 5—is advancing on a similar exponential curve.

In the DARPA Grand Challenge of 2004 for autonomous vehicles, no self-driving car came close to finishing the course. Just eleven years later, self-driving cars are legal in more than a dozen states and are a common sight on the streets of the Bay Area of San Francisco. Incidentally, three teams, with three different designs, completed DARPA's 2015 Challenge course.

In voice recognition, robots are already close to attaining the capabilities of C-3PO. Apple, Amazon, and Google do decent jobs of translating speech to text, even in noisy environments. Their voice-recognition systems struggle with accents, words difficult to pronounce, and colloquial abbreviations, but they are, in the main, quite serviceable. Though no A.I. bot has passed the Turing Test—the gold standard of A.I., whereby humans are unable to distinguish a human from a robot in conversation—the machines are getting closer. Siri and her compatriots will soon be able to converse with you in complex, human-like interactions.

Still, machines have yet to crack voice recognition in more complicated, multi-voice environments, where the task involves recognizing the voice communications of several humans simultaneously in a loud environment. This is a far more difficult task than single-voice recognition and illustrates the sophistication of processing in our brains.

The computational demands of these difficult tasks would seem to be insurmountable obstacles to robots whose development is linear. But, as I explained earlier, by 2023 your iPhone will have computational power equivalent to that of a human brain. It is becoming clear that, given A.I.'s continuing exponential progress, robot builders' and A.I. coders' present computational obstacles are on the verge of becoming irrelevant.

The rise of machine learning, too, heralds a generation of robots that can learn through doing and that will become smarter as they spend more time with us. Google is on the verge of completing real-time text- and voice-translation software, built, in part, with human input through Google Translate. Google's DeepMind system, which beat the world's leading Go player in 2016, learned to play this millennia-old board game, orders of magnitude more complicated than chess, by watching humans play Go.[3] Even more fascinating, DeepMind surprised human Go experts with moves that, at first glance, made no sense but ultimately proved innovative. The humans taught the robot not just to play like a human but how to think for itself in novel ways. Though not passage of a Turing Test, this is a clear sign of emergent intelligence, distinct from human instruction.

For all of these reasons, I expect that a robot maid—a robot like Rosie—will be able to clean up after me by 2025. Robots will soon become sure-footed; and a robot will, rather than merely open a door, succeed in opening it while holding a bag of groceries and ensuring that the dog doesn't escape. When I buy Rosie, I may have to show her around the house, but she'll quickly learn what I need, where my washer and dryer are located, and how to navigate around and clean the bathroom. And I expect that she will be as witty and lovable as she was on TV. No, she won't have the artificial general intelligence that will make her seem human, but she will be able to have fun conversations with us.

In fact, a very limited version of Rosie can be found at hospitals around the country. Her name is Tug, and she is produced by Aethon Inc. of Pittsburgh. Tug performs the most essential duties of today's hospital orderly, such as delivering medications and equipment to different floors. Tug costs considerably less than the orderly position she replaces.

But Tug doesn't clean up rooms or do anything more

complex than navigate the hospital corridors. The idea of robots' replacing humans wholesale and quickly is unrealistic. Rather, the robots will replace humans piecemeal in performing tasks, through specializations.

In this fashion, the robots will gradually, task by task, assume the jobs of humans in manufacturing plants, in grocery stores, in pharmacies. Hospitals rely on A.I.-driven systems in their pharmacies right now to spot potential problems due to conflicting medicines. I can envisage the job of pharmacist being completely automated. Further down the economic food chain, McDonald's is in the process of rolling out automated order-taking at its counters. This could be matched by an automated engine to cook hamburger and fries. One of these already exists. It's from a venture-backed company called Momentum Machines and can make a hamburger every ten seconds. That may sound ominous; yes, robots may eat our jobs. But, in the rapidly aging developed world, we may need robots to take care of our aging populations and maintain economic stability in the face of staggering demographic change that will leave us collectively with far more work to do than workers to do it can accomplish.

How Robots May Save the World

According to the U.S. Census Bureau between 2020 and 2030, for the first time in human history, the global population of people older than sixty-five years will eclipse the population of people under the age of five.[4] The trend is moving faster in developed countries but is also accelerating even in less-developed countries. According to the Population Reference Bureau, the proportion of people aged sixty-five or more in less-developed countries has increased by 50 percent since 1950, from 4 percent to 6 percent.[5] In more-developed countries, the proportion of people aged sixty-five or more, which was 8 percent in 1950, increased to 16 percent in 2014 and should rise to a record 26 percent by 2050. In those developed countries, the rate of childbirth is declining.

This represents a trend toward an aging populace, with fewer and fewer people supporting, both through health care and through taxation, the ballooning ranks of the retired and aged. In a country such as Japan, the ratio of retired people to working adults will continue to soar beyond its current record. That could tax the country as nothing has before: caring for an aging population could overwhelm all other economic and political priorities. For this reason, the Japanese government has embraced the concept of robotic caregivers for the elderly.

Japan's robot solution to its unavoidable aging crisis calls up a future that makes many uneasy, in which the aged are wholly reliant on robots and in which people no longer invest significant energy in caring for their parents. This sits uncomfortably with deeply held societal ideals about filial piety and blood duty to take care of those who protected and fed us when we were small. But such may be the tradeoffs required to sustain an economy in which fewer and fewer people of working age are deployed to productive occupations in society. At its core, too, this touches on the highly emotional topic of whether we are better off ceding nearly all work to the robots.

The robots may become not only a saving grace for an aging developed world—but also our best friends.

Should Robots Kill People?

Japan may favor robots to protect its elderly and preserve its economy, but a far more contentious discussion is under way right now with tremendous implications for humanity, concerning use of robots for destructive purposes. The debate concerns whether we should allow robots powered by A.I. to kill people autonomously. More than 20,000 people signed an open letter in July 2015 that called for a worldwide ban on autonomous killing machines. A thousand of these signatories were A.I. researchers and technologists, including Elon Musk, Stephen Hawking, and Steve Wozniak.[6] Their logic was simple: that once development begins of military robots enabled to autonomously kill humans, the technology will follow all technology cost and capability curves; that, in the not-so-distant future, A.I. killing machines will therefore become commodity items, easy to purchase and available to every dictator, paramilitary group, and terrorist cell. Also, of course, despotic (or even wayward democratic) governments could use these machines to control and cow their populations.

This is a point that almost everyone can agree on, with the exception of a handful of people in the military establishment. Even Ray Kurzweil, who is about as pro-robot as you can find, is staunchly opposed to programming robots to kill people without asking permission from a human controller. Such programming would, he believes, be a moral violation. Other critics, such as AJung Moon, cofounder of the Open Roboethics initiative, fear that allowing autonomous lethal force will tip us down the slippery slope toward a world in which the machines could act autonomously beyond the intent programmed into them.[7] And, as DeepMind demonstrated on the Go board, robots made smart enough will likely have minds of their own, within at least the rules and environment they have mastered.

The military supporters of autonomous lethal force argue that robots in the battlefield might prove to be far more moral than their human counterparts. A robot programmed not to shoot women and children would not freak out under the pressure of battle. There would have been no Mai Lai massacre if the robots had been in charge, they say. Furthermore, they argue, programmatic logic has an admirable ability to reduce the core moral issue down to binary decisions. For example, a robot might decide in a second that it is indeed better to save the lives of a school bus full of children over the life of a single driver who has fallen asleep at the wheel.

These lines of thought are interesting and not wholly unwarranted. Are humans more moral if they can program robots to avoid the weaknesses of the human psyche and the emotional frailty that can cause even the most experienced military man to temporarily lose his sense of reason and morality in the heat of battle? Where it is hard to discern whether the opponent follows any moral compass, such as in the case of ISIS, is it better to rely on the cold logic of the robot warrior rather than on an emotional human being? What if a non-state terrorist organization devises lethal robots that have a battlefield advantage? Is that a risk that we are willing to take by developing them?

I have a cynical view on this: I do not think the public really cares much about whether robots will be allowed to kill people, the notion seeming too abstract. The American public has never taken much interest in whether drones should be equipped for autonomous kill shots. In fact, the public has taken little interest in the question of robots being used to kill people even in the United States. In Dallas, police used a bomb carried on a robot to kill Micah Brown, the shooter who had allegedly killed seven officers at a protest rally.[8] Few questioned this use of force. And the first autonomous robots' use on the battlefield would likely be far away, as the battlefields on which drones first killed were, in Afghanistan and Pakistan.

The Open Roboethics initiative is advocating an outright ban on autonomous lethal robots, a call echoed by nearly every civil rights organization and by many politicians. The issue will play out over the next few years. It will be interesting to see not only what final decision comes from world governance bodies such as the United Nations but also the decision of the U.S. military establishment and its willingness to sign an international accord on the matter. (The United States is a consistent holdout on restrictive treaties regarding military technology, a realm in which the nation holds a clear global advantage.)

Do the Benefits Outweigh the Risks?

So now we come down to the question at hand. Do the benefits of robots outweigh the risks? If so, how do we mitigate the risks? Stopping altogether the penetration of robots into society and the world is by now a lost cause. Tug is not going back in the box. Google cars—robots that drive vehicles for us—are here and will probably not be stopped. Tesla cars with autopilot capabilities have already driven millions of miles on our highways. And as A.I.-endowed robots advance, inevitably emergent capabilities will result in things we have not expected. The extreme risk is apocalyptic: the robots become smarter than we are and take over the world, rendering humans powerless on their own planet.

An equally troubling but less existential, and more realistic, risk is that the robots increasingly deprive us of our jobs. Some researchers, such as Erik Brynjolfsson and Andrew McAfee of the Massachusetts Institute of Technology, see the automatons inevitably gobbling up more and more meaningful slices of our work.[9]

Oxford University researchers Carl Benedikt Frey and Michael A. Osborne caused a tremendous stir in September 2013, when they asserted in a seminal paper that A.I. would put 47 percent of current U.S. employment "at risk."[10] The paper, "The Future of Employment," is a rigorous and detailed historical review of research on the effect of technology innovation upon labor markets and employment. In a recent research paper, McKinsey & Company found that "only about 5 percent of occupations could be fully automated by adapting current technology. However, today's technologies could automate 45 percent of the activities people are paid to perform across all occupations. What's more, about 60 percent of all occupations could see 30 percent or more of their work activities automated."[11]

The report also notes that the mere ability to automate work doesn't make it a sensible thing to do. As long as $10-per-hour cooks are cheaper than Momentum Machines on fast-food lines, it's unlikely that food-service jobs will succumb to automation.

The alternative extreme—no robots—is simply not realistic. The giant bubble of aging people could overwhelm most of the developed world, as well as many developing countries, such as China. Self-driving cars will save tens of millions of lives over the next decades. More agile and intelligent robots will take over the most dangerous human tasks and jobs such as mining, firefighting, search and rescue, and inspecting tall buildings and communications towers.

To put this in perspective, it seems we want robots that can do everything humans have trouble doing well as long as the robots don't capture our most unique capabilities or become much smarter than we are. Perhaps this is a good thing. A robotic caregiver, which on the surface may seem a heartless option, is compassionate compared with providing no caregiver or placing a child under harsh financial pressures. And, taking this argument further: perhaps the robots and the economic gains they allow will take our jobs but also provide us humans with huge gains in free time to pursue our passions.

To me, the crux of this matter will be maintaining the ability of humans to understand robots and stop them from going too far. Google is looking at building in a kill switch on its A.I. systems.[12] Other researchers are developing tools to visualize the otherwise impenetrable code in machine-generated algorithms built using Deep Learning systems. So the question that we must always be able to answer in the affirmative is whether we can stop it. With both A.I. and robotics, we must design all systems with this key consideration in mind, even if that reduces the capabilities and emergent properties of those systems and robots.

Will All Benefit Equally?

On the question of whether the technology will benefit everyone equally by the availability of general-purpose robots, the answer is no: the rich will surely benefit more than the poor—because they will get the latest and greatest robots first. Note the differences between the smartphones we use here in the United States and those common in India and China. Our devices often cost more than $600, and theirs are as cheap as $40. We have the fastest processors, the longest-lasting batteries, and the best screens. Their devices are two to three years behind ours in features and function. But, frankly, I don't see this as a problem, because effectively the rich are subsidizing the poor, paying for the technology advances. This was the same strategy that Elon Musk used in building the Tesla: with the Roadster and Model S, he targeted the high-end market and had people like me pay for the advances that will lead to a more affordable Model 3.

Fans of the TV series *The Jetsons* may also recall the first episode, when Jane Jetson bought Rosie because she was an "old demonstrator model with a lot of mileage." She was the only robot that this middle-class family of the future could afford. Rosie may not have had the latest features, but she provided tremendous benefit. That is the future we are looking at with robots—one in which everyone will eventually benefit, although some may get the advanced models earlier than others.

The worrisome issue when considering social equity is, as I discussed earlier, the disruption that robots will create in employment. This will have serious consequences unless we develop the safety nets, retraining schemes, and social structures to deal with an era of abundance and joblessness. We need to help people adapt to a new social order in which our status isn't just based on the job we do, but based also—and perhaps more—on the contribution we make.

And, no matter what governments do, they will not be able to prevent automation, because it is a key to economic growth. There is no way we can sugarcoat the impact of technology on employment; we simply need to prepare for it. We need to learn about where technology is heading, to understand its impact, and to cushion the blow to those who will feel its negative effects most.

Does the Technology Foster Autonomy Rather Than Dependence?

Will we be dependent on our robots? To some extent, we already are, on the primitive ones: our cars, elevators, dishwashers, and practically everything else that runs on electricity. We will surely have an option of not using them, but that wouldn't make for an easy life. I can tell you that it is more than gadgets with electric cords; without my smartphone and Internet access, I feel lost. Why would our dependence upon the robots that serve us and become our friends and companions be any different?

• 9 •

Security and Privacy in an Era of Ubiquitous Connectivity

In an episode of the popular TV series *Homeland,* Vice President William Walden is killed by a terrorist who hacked into Walden's heart pacemaker. The hacker raises Walden's heart rate, pushing him into a serious, inevitable cardiac arrest. Walden's pacemaker had been connected to the Internet so that his doctors could monitor his health. That was the fatal mistake. Viewers watched in shock and disbelief, but this assassination plot seemingly out of science fiction was actually not that far-fetched.

These days, many complicated, critically important medical devices include onboard computers and wireless connectivity. Insulin pumps, glucose monitors, and defibrillators have all joined the Internet of Things. Every year at security conferences, hackers are demonstrating new ways to compromise the devices we rely on to keep us alive. Former Vice President Dick Cheney famously asked his doctors to disable the wireless connectivity of the pacemaker embedded in his chest. "It seemed to me to be a bad idea for the vice president to have a device that maybe somebody on a rope line or in the next hotel room or downstairs might be able to get into—hack into," Cheney's cardiologist, Jonathan Reiner of George Washington University Hospital in Washington, D.C., told *60 Minutes* in an interview in October 2013.[1]

We will live simultaneously in an age of wondrous technical marvels and one of perpetual insecurity, and such threats will become more common. Those individuals and groups who wish to do us harm are more empowered than at any time in the past. Blackmail using purloined personal data will skyrocket. We will begin to understand the disadvantages of having devices always collecting information and companies offering products and services for free. Cybersecurity will move from an abstract threat to an issue of personal safety that will matter to us all.

So be ready for a rough twenty years ahead. But there is some good news. The cybersecurity industry is already responding, and technologies that could mitigate these threats are already under development. The next generation of security experts is stepping up to the challenges and creating innovative solutions. Governments, corporations, and entrepreneurs everywhere understand the benefits of solving these issues and are racing ahead with novel approaches and breakthrough methods. Each advance we make will come with setbacks, but we will work through those as we go. The question is what will we lose in the process?

Citizens Caught in the Cyber Crossfire

The ability to access nearly all of the world's information from an affordable personal supercomputer in your pocket has unquestionably brought benefits. We can reach loved ones at a moment's notice, access a rapidly growing list of services instantly, and learn almost anything we want from anywhere. It's not just the rich who are benefiting; it is arguable that the greatest gains are being made by the global poor, who can now communicate, collaborate, and bypass some of the institutional barriers that have held them back.

As high-speed, ubiquitous connectivity among all manner of devices binds us more tightly to technology and to the Internet, a crucial and frightening mega-trend for the next two decades is that cyber security will become a more important domestic-security issue. In 2007, the Stuxnet computer worm sent costly and critically important centrifuges spinning wildly out of control at Natanz, a secret uranium-enrichment facility in Iran.[2] In a matter of months, American and Israeli security forces were able to remotely destroy 1,000 of the 5,000 centrifuges Iran had spinning at the time to purify uranium. The government program behind the virus, code-named "Olympic Games," was developed during the Bush and Obama Administrations.

Stuxnet was the first major publicly reported governmental cyber attack on industrial facilities of another nation.

Then, in 2015, American intelligence services suffered their worst defeat in modern history, at the hands of intruders believed to be from China. The Office of Personnel Management, the government agency responsible for vetting and managing employees, suffered a catastrophic data breach that exposed its full records of more 21.5 million employees, dating back almost thirty years.[3] The stolen data included more than five million sets of fingerprints, which can never be changed. Even worse, the personal details and secrets of more than four million security-clearance holders were also leaked, forever changing the country's ability to conduct espionage abroad.

In 2016, hackers, allegedly Russian, compromised e-mail servers of Democratic Party officials and tried to use this information to undermine trust in the U.S. electoral process.

The next major geopolitical crisis will involve not only electronic countermeasures against enemy missiles and communication systems but also attacks over IP networks to cripple or destroy civilian infrastructure. Our personal information and security will be collateral damage in the continuing battle between nations for control.

As we rush headlong into the Internet of Things and connect willy-nilly everything that can be connected, we expose the soft underbelly of our technological systems. Identity theft has intensified significantly in the past two decades, but the public remains in the dark about its growth in sophistication behind the scenes. The next two decades will mark a change from inconvenience to real harm. As we read more about thefts of celebrities' nude photos and exposure of people's e-mail, hacking will become something all of us worry a lot more about.

Loss of financial identity is one thing. What is coming

now is much uglier—and personal. It is far more difficult to recover from a leak such as the attack on Ashley Madison.[4] The publication of e-mail addresses of alleged customers of the online adultery-facilitation service exposed millions of people to ridicule and marked them with a virtual scarlet letter. These suspected cheaters are now searchable in a number of databases, forever. It even drove some to suicide. Data breaches don't take account of nuance; the devastation of their personal and social lives will be unmitigated by whether a couple was going through a rough patch or whether somebody was just looking with no intention to actually commit infidelity.

It's not just the things we say or do but also the information that is collected about us that makes up our identity and reputation now. On a typical day as you drive home, cameras mounted on top of police cars and road signs are using automated license-plate recognition technology to make a database of virtually all of your car's movements. Surveillance cameras on buildings and at traffic stops are constantly snapping pictures and recording video of you everywhere you go. As you pull into your driveway, your home automation system makes a record of exactly when you arrived; to deliver the perfect temperature, your Nest thermostat tracks your movements across the house. The cameras and microphones on your Smart TV listen in to all of your conversations, waiting for you to issue the TV with a command. And that's all before you launch your web browser.

All Your Weaknesses, in One Place

As we move toward a connected system and toward having our lives tied to our cloud services, we create more and more single points of failure that can grind our existence to a halt. When then *WIRED* magazine reporter (and now BuzzFeed tech editor) Mat Honan had all of his digital belongings deleted, the hackers didn't use some cutting-edge technology or brute force to make their way in. Instead, they used social engineering to trick Apple and Amazon customer-support personnel into giving control of Honan's account to a stranger.

Writes Honan, "In the space of one hour, my entire digital life was destroyed. First my Google account was taken over, then deleted. Next my Twitter account was compromised, and used as a platform to broadcast racist and homophobic messages. And worst of all, my AppleID account was broken into, and my hackers used it to remotely erase all of the data on my iPhone, iPad, and MacBook."[5]

Some of Honan's lost items were pictures of his young child that he had forgotten to back up. They're now lost for good.

Honan was targeted because he had a highly coveted three-letter Twitter handle. There will be many, many more Mat Honans in the next few years, as whole Dropboxes, Google Clouds, and iCloud accounts of many people will be wiped out (at least temporarily) by hackers who turn their victims' lives upside down, spoil their reputations, and extort money or promises from them. Our own unsuspecting behavior on social media offers only additional surface area for attack. We post pictures of the cars we drive, talk about the places we eat at, publicly reveal our work histories and our personal networks, and publish links to articles on publications we subscribe to without giving a second thought to how that information could be later used to hijack our identities.

Centralized databases and stores of personal information can have risks beyond the financial and social. Medical identify theft is growing rapidly, in which someone can use a stolen social-security number to receive health care under your name and pay for it with your insurance. Unfortunately, you may be left paying the doctor's bill. And as we connect all of our electronic medical records systems and pipe them into larger A.I. systems such as IBM's Watson, false data about our health becomes harder to expunge from our permanent record.

This tampering could result in poor diagnoses and potentially hazardous treatments or care. Imagine that someone using your insurance fills out a doctor's office standard form on allergies and claims to have none—and you have a very dangerous drug allergy. If you are in a car accident and that drug is a standard course of treatment for your injury, the latest record might show no allergy. Unconscious and unable to correct the record, you experience a dangerous allergic reaction: tragic, and eminently preventable.

The Race to Make Security Accessible

As it stands today, there are plenty of tools you can deploy to get a very high level of security and protection. The problem is that they are nearly impossible for the average person to use: technically complicated, requiring great expertise, and with awful user interfaces, especially in comparison with their less secure alternatives. Security technologies need to become more user friendly.

There has been some progress in that direction. Secure personal clouds offer a viable alternative to using services like Dropbox, and they can lay the groundwork for a system where we are able to charge for access to our data rather than having it taken from us. Tools are on their way that can help us control our digital footprint and manage who is able to shape it. Companies too are finally implementing default settings that are supportive of how users actually behave, instead of tricking people into sharing more information. Facebook has put in place some of the best systems for blocking social-engineering attacks, by examining whether the hacker asking for your password is likely to be you or someone else, based on a host of key signals such as location, type of computer, time of day, browser version, and more.

On the extreme dark side of the security and privacy discussion for the future is the inevitable decoding of our DNA, the inevitable capture of our biometrics whether we like it or not (facial recognition, voice, gait, fingerprints), and the capture of every moment of our daily lives. We are going to need to think deeply about how much we value our individual privacy.

A Difficult Balance

Transparency, detection, and accountability are the necessary antidotes to security risks. Companies need to build systems with the assumption that they will be hacked. They need to develop technologies that notify us when we've been compromised and take automatic actions to block attackers. They must design systems to be distributed and resilient, such as blockchain technology, which can help prevent tampering and information leakage.

With regard to privacy, we have yet to reach a consensus on what is acceptable. We all make choices about what we put on line, but much of what is collected about us is out of our control. The actual value of privacy is up to citizens and governments of the world to decide. Perhaps we need a blanket ban on covert capture of facial-recognition identification. Maybe we need to mandate that any system that scans faces in public places must be clearly marked and announced. Perhaps we need to reform liability laws to make developers and manufacturers of our devices take our security more seriously. Or maybe we need an amendment in the Constitution that says we own our data, so that we are finally on a level playing field legally with the technology companies presently able to swipe them from us and use them against us with impunity.

It is the job of governments to enact laws that protect

the public, but we must tell our policy makers what we want. As I have said before, laws are codified ethics; our political leaders are supposed to do what we say, to implement policies that we have reached a social consensus on. The Europeans, for example, are tightening regulations on U.S. technology companies by requiring them to adhere to stricter standards and to store data locally rather than across borders. But this is little more than a Band-Aid.

There is another way of forcing technology companies to be more prudent with our data. Insurance companies selling cyber insurance are raising rates; and applicants, in order to receive coverage, often necessary for doing business, must undergo security audits. Putting a higher price on personal privacy and making its breach a more acute financial risk to businesses would probably force companies to think a lot more about how they are securing your data. As the problems of extortion, ID theft, and hacking grow more acute in the short term and as the value of privacy enters the public's consciousness, it becomes easier to get such measures passed. The most effective time to convince people of the need to take corrective measures is unfortunately right after they have been compromised—when they most clearly understand the consequences of inaction.

So we can expect our identities to be stolen; we can expect extortion attempts; we can expect attempts at scary industrial hacks. But the worst problems of the last generation of technology are often easily solved by the first generations of the next wave of technology—until they create their own issues that need solving.

Do the Benefits Outweigh the Risks?

Increasingly ubiquitous digital-information capture clearly represents a tremendous risk to each and every one of us. These practices are difficult to track: it's increasingly hard to follow who knows what about us, and where they learned it. The convenience of our digital existences, from online photos to social networks to online document storage, is undeniable and likely irreversible. So do the benefits outweigh the risks?

I have very mixed feelings about whether the risks we face are worth the benefits we receive from putting so much of our data on line so unprotected. Because the system governing use of data on line is not a system at all but an ad hoc jumble of commercial relationships with thin legal protections and even thinner real-world protections, for me the conveniences of one-click online orders and automated log-ins to websites courtesy of Facebook are thin gruel compared with the larger risks we face. The big problem is that users (meaning you and I) have only two alternatives: opt in, or opt out.

That is a choice we should not have to make. Newer practices of managing sensitive data can put users in charge or, alternatively, collect only the data necessary to perform the task at hand. We need a radical shift in how we think about data collection, centering system design on users' data management and their privacy rights rather than layering them on as an afterthought. Users will vote with their online presence. Noted futurist and author

Kevin Kelly observes in his book *The Inevitable* that "vanity trumps privacy"—that we are willing to give incredibly revealing details about ourselves in exchange for social validation: "They'll take transparent personalized sharing. . . . If today's social media has taught us anything about ourselves as a species, it is that the human impulse to share overwhelms the human impulse for privacy."[6]

This has been true, in part, because the costs of losing control of our data are hidden and hard to understand. But as identity theft reaches epic proportions and very little of our personal information is left untouched by credit bureaus and malicious online thieves alike, I predict that issues of data security will become far more acute to far more people, making security and privacy unavoidable issues. Then privacy will trump vanity.

In a nutshell, in the present state of affairs, I am not at all convinced that sacrificing our security and privacy for online convenience is worth the price. More accurately, I resent that we are even forced to make such a choice, given the badly structured and poorly policed governing systems for online security and privacy that we presently endure. So far, it's not worth sharing all your data on line and trusting that nothing bad will happen. If you must share them, I'd recommend mitigating your risk by managing and understanding how data are being used. Yes, it's an almost impossible task right now. And it's not going to become more straightforward until we push for it, which is why learning about these technologies and their effects is so important.

10

The Drones Are Coming

You have probably had to pop out to the grocery store to pick up something you needed for a dinner party. Or maybe you've dashed to the pharmacy to get a prescription refill before you took a long trip. By the early 2020s, small drones will do that, and a whole lot more, for you.

Companies such as Amazon and Google have long been planning drone-delivery services, but the first authorized commercial delivery in the United States happened in July 2016, when a 7-Eleven delivered Slurpees, a chicken sandwich, donuts, hot coffee, and candy to a customer in Reno, Nevada.[1] In the United Kingdom, an enterprising Domino's franchisee had made headlines by using a drone copter for deliveries in June 2013. Hundreds of companies delivering by drone are starting up all over the world. Venture-capital firm Kleiner Perkins estimates that there were 4.3 million shipments of drones in 2015 and that the market is growing by 167 percent per year.[2]

Not since the automobile has a transportation technology spurred such enthusiastic entrepreneurial activity. The barrier to entry into the business of building drones is exceptionally low. Commodity kits compete with commercial models, and Arduino circuit boards and open-source software make it easy for motivated coders and hackers to tailor drones to exacting functions in arcane and lucrative fields. Just a decade after the military began using drones in earnest as remote-controlled killing machines, the same technology is available to everyone (but not to hunt down terrorists).

Drones are also known as Unmanned Aircraft Systems (UAS) and Unmanned Aerial Vehicles (UAVs). Their evolution is an excellent example of how exponential technology development works. People have been using radio signals to fly aircraft remotely for more than fifty years. The problem with piloting these devices was keeping them stable and preventing them from crashing when there was any wind turbulence. That necessitated piloting by very highly skilled operators. Drones had to carry big cameras and transponders in order to transmit images and data to the operator. Now, cheap, powerful, light computers and sensors can do that job, lowering drone-manufacture costs and enabling exponentially faster processing in the payloads that drones' propulsion systems can comfortably send aloft.

What has really changed the game is the autopilot. In existence for military and commercial flights for many decades, autopilot software that worked well and was available to everyone came along only a few years ago. In part, it was not considered necessary, because the FAA does not allow over-the-horizon drone operations in public space. But the agency appears to be ready to start permitting it, and drone builders have been experimenting with private over-the-horizon flight for years. The units now entering testing will follow a route set by map pins, and can automatically land or return to a charging station as necessary. You can buy a Parrot AR.Drone, for example, for about $200 on Amazon. This quadcopter transmits 720p high-definition streaming video to the iPad or smartphone controlling it. It is equipped with a three-axis accelerometer, gyroscope, and magnetometer, as well as pressure and ultrasound sensors. Two or three decades ago, such sensors would have cost hundreds of thousands of dollars and weighed tens of pounds. They are, essentially, military-grade hardware and technology. In China it is now possible to buy drones that are more or less the equivalent of U.S. military models for only a few thousand U.S. dollars.

We are entering what Chris Anderson, the founder of a company called 3D Robotics, calls the "Drone Age." Anderson was one of the early entrants in the DIY-drone surge and is one of the leading advocates of widespread drone adoption. "It's safe to say that drones are the first technology in history where the toy industry and hobbyists are beating the military-industrial complex at its own game," wrote Anderson in a 2012 piece in *WIRED*.[3]

This is both good and bad. Drones' ability to travel directly to their destination on uncrowded flight paths will enable them to replace all manner of terrestrial shipping. In cities and suburbs, drones will replace delivery vehicles. This will reduce urban congestion and possibly carbon emissions, and save money and trips to the emergency room (car accidents kill, you know). Presently, when you order a pizza, delivery of a pound of flour with some toppings and tomato sauce involves sending to your driveway a human being in a two-ton vehicle spewing carbon. A small drone weighing a few pounds will perform the delivery better, day or night, rain or shine.

Drones can also perform jobs hazardous for humans to perform, such as inspecting roofs, cellphone towers, and bridges. In drought zones such as California, drones can perform round-the-clock fire spotting, with nearly 100 percent coverage of the entire state, to quickly locate wildfires.

Because drones are so cheap and are getting cheaper by the month, they hold tremendous potential in the developing world to provide the same aerial services the West will soon enjoy. That may allow these comparatively poor parts of the planet to leap forward into a more modern, more efficient era. In sub-Saharan Africa and parts of Asia, for example, such a service could be critical, because unreliable transport networks can cause the supply of spare parts for farm equipment or medical equipment to take weeks or months.

It's already beginning to happen. In Malawi, UNICEF is looking to start testing drone delivery of medical samples to remote regions of the country.[4] Not confined to the developing world, precisely this service was tested in impoverished rural West Virginia. In July 2015, a hexacopter drone operated by the Australian startup Flirtey was deployed to deliver boxes of prescription medicines to a remote pop-up field clinic in rural Wise County, West Virginia.[5] Moving the boxes by drone rather than by traditional means allowed for much faster resupply of critical medicines and, in general, allowed for same-day delivery of necessary items.

That U.S. doctors would choose to use drones to deliver medicines in rural areas hints at another possible profound equalizing effect of drones. The United States has rapidly urbanized. Jobs and resources have flowed to urban and densely populated areas, and parts of rural America have steadily hollowed out, becoming poorer. In combination with decaying infrastructure, this has created a country of urban and suburban haves and rural have-nots. Drones could boost living standards in rural America as they boost convenience in urban America. If a drone could deliver your groceries in West Virginia at a low cost, saving you an hour's drive to Walmart and the cost of gasoline (not to mention your time), that's a real improvement in the standard of living. If a small factory needed spare parts from the nearest distributor, then a direct flight by a drone would trump a FedEx delivery both in cost and in speed.

There are also tremendous applications of drones to agriculture, such as monitoring the growth of weeds and crops, spraying insecticides when needed, and tracking soil hydration to adjust watering. Drones can enable a process known as precision agriculture, which optimizes the use of resources and reduces the amount of runoff that could flow into nearby rivers and streams.

The Darker Side to Drones

In June 2015, a hobby drone flew into the path of an air tanker fighting a 17,000-acre forest fire in California, causing the state fire agency Cal Fire to ground all nearby air operations for the evening.[6] The incidence of such dangerous drone near-misses is on the rise. On Sunday, August 16, of the same year, the FAA recorded a total of twelve episodes involving rogue drones, sixteen endangering manned aircraft, in five different states. On that day, flabbergasted pilots reported that two large commercial jets had near misses by drones above Los Angeles International Airport. The FAA has tallied hundreds of instances in which drones have entered restricted airspace or nearly struck other aircraft.

Bad drone behavior is not confined to unintentional transgressions. Criminals have embraced drones. They have taken to the air to smuggle illicit drugs into a prison in Ohio. Mexican drug runners used a drone to carry twenty-eight pounds of heroin across the U.S. border in August 2015.[7] And an eighteen-year-old mechanical-engineering student modified a drone to carry a handgun that could be fired remotely. (He was arrested for this stunt.)[8]

We know too that drones have been adopted into the arsenals of anti-Western groups. The militant Islamist group Hezbollah is building a massive drone air force to take to the skies against Israel. Hezbollah has already dispatched drones over the border on several occasions, flying very close to key infrastructure. In October 2016, an ISIS drone blew up and took the lives of two Kurdish fighters that were trying to stop it from doing surveillance.[9]

For Hezbollah, ISIS, and other non-state actors, the drone could become a great equalizer, a mechanism for delivering "suicide" bombs that requires no recruits and no explosive vests. Though countries are working on drone defenses, ranging from shooting down relatively slow-moving drones to jamming their GPS signals, it's unclear whether any country actually has a viable defense against swarms of drones bearing explosives. And other drone-mounted weapons will doubtless follow, such as machine guns and poison gas.

Do the Benefits Outweigh the Risks?

The overall desirability of drones really depends on how much abuse we see and how rapidly we develop defenses against such abuses. Unlike the loss of personal data protection and privacy, we as a society could conceivably opt out of drones by banning their sale, restricting their operation, and constructing electronic countermeasures to their remote control. And we could develop technologies that incapacitate them in certain areas.

Drones offer a healthier balance between promise and peril than most of the technologies in this book, though. And there are things we can do to mitigate the risks.

To start with, there needs to be a core technology framework for collision avoidance. Though this is no trivial problem, it is looking increasingly soluble. Self-driving cars address many of the same issues, and do so in an even more crowded and dangerous landscape, filled with unpredictable humans doing silly things such as texting while driving. The next generation of self-driving cars' laser sensors will be embedded in the vehicles' chassis. Drones might not even need that much sophistication in order to avoid collisions: a simple system in which every drone in operation emits a signal to alert other drones to its proximity may suffice (except in avoiding the occasional bird strike).

The next step would be to build a system of air-traffic control for drones. It would need to be automated and to include safety measures such as emergency kill-switches to bring down a drone that is malfunctioning or poses a danger. We would need to specify city air corridors dedicated to drones and to confine the drones to them.

We also need to build private and commercial air-defense systems, such as the military is developing, to shield our schools, homes, and businesses from drone surveillance and attack; eventually, it might be something like the invisible shields depicted in *Star Trek*.

All of these are technically feasible. Beyond this, we need to debate what is socially acceptable, and to create legal frameworks. Should the cameras of delivery drones be recording and saving all video footage as they enter the airspace of a customer's home? For that matter, should drones be allowed to fly over private property at all, or should they be limited to public roads between droneports? Should we have the right to shoot down unauthorized drones on our property? If the Second Amendment grants the right of gun ownership to individuals for self-defense, then does it allow them to fly their own defensive drones?

The FAA implemented new regulations in August 2016 for the commercial use of drones in the United States and is working on updating them as it learns drones' uses and needs.[10] It stipulated that drones had to remain in visual line of sight of the pilot; that operations had to be in daylight hours; that operators must be at least sixteen years of age; that the groundspeed maximum would be 100 mph and altitude maximum, 400 feet; and that pilots must obtain certification. These rules governed tasks such as surveying, real-estate photography, and site inspections and did not apply to drone delivery operations—because those use autonomous technology, rather than human pilots, to guide them.

State legislatures across the country are also debating how human-piloted drones should be regulated. They are trying to address the needs of law enforcement and to answer questions that their constituents are asking. Drones can be used for hobby and recreation, and for hunting game. According to the National Conference of State Legislatures, as of 2016, thirty-two states had enacted laws addressing issues with unmanned aircraft systems, and an additional five states had adopted relevant resolutions.[11]

That the FAA and the states are actively researching the issues and listening to business and the public is a good sign. The industry will develop a lot faster, and there will be better protections, if there are clear and sensible regulations.

Does the Technology Foster Autonomy Rather Than Dependence?

In the case of drones, these are both relatively easy choices. Everyone benefits from a reduction in the cost of delivery and the surmounting of obstacles to get goods to where they are needed. The poor gain as much as do the rich. After all, drones are becoming so cheap that almost everyone will be able to afford them. And when it comes to autonomy, we have a clear choice: if we don't want Starbucks droning our morning latte, we can always drive to the store to pick it up. Or we can have our self-driving car take us there.

These considerations over drones illustrate well a point that I have been trying to make: the imperativeness that we, the public, learn about advancing technologies, decide what we consider to be ethical and acceptable, and tell our policy makers what regulations to enact. It is the key to building a *Star Trek* future.

11

Designer Genes, the Bacteria in Our Guts, and Precision Medicine

In the near future, we will routinely have our genetic material analyzed; late in the next decade, we will be able to download and "print" at home medicines, tissues, and bacteria custom designed to suit our DNA and keep us healthy. In short, we will all be biohackers and amateur geneticists, able to understand how our genes work and how to fix them. That's because these technologies are moving along the exponential technology curve.

Scientists published the first draft analysis of the human genome in 2001. The effort to sequence a human genome was a long and costly one. Started by the government-funded Human Genome Project and later augmented by Celera Genomics and its noted scientist CEO, Craig Venter, the sequencing spanned more than a decade and cost nearly $3 billion. Today, numerous companies are able to completely sequence your DNA for around $1,000, in less than three days. There are even venture-backed companies, such as 23andMe, that sequence parts of human DNA for consumers, without any doctor participation or prescription, for as little as $199.

We can expect the price of DNA sequencing to fall to

the cost of a regular blood test in the early 2020s and, shortly thereafter, to cost practically nothing. Again, what makes this possible is that the computers that sequence DNA are becoming faster and more powerful in parallel with development of the microprocessors that power them, which double in speed and halve in price every eighteen to twenty-four months. Ultimately, someone will build a sensor-packed dongle and a smartphone application that can do it in the field, in seconds: prick your finger; parse your DNA; done.

By the mid-2020s, sequencing DNA will probably become part of a normal health panel. Your genome will be part of what doctors look at to determine treatments and potential risks. This will be more accurate than any other tests.

A team of scientists published a study in *The New England Journal of Medicine*, in March 2014, documenting that fetal DNA testing was ten times better in predicting cases of Down syndrome than the standard blood test and ultrasound screening were, and five times better in predicting another disorder, trisomy 18, a mutation arising from an error in cell division that, in the early months and years of life, has more potentially life-threatening complications.[1]

At New York's Memorial Sloan Kettering Hospital, one of the world's leading cancer-treatment centers, scientists have been pioneering DNA tests that allow doctors to quickly find out whether a patient's tumor carries clinically useful mutations that make cancers vulnerable to particular drugs and to match individual patients with available therapies or clinical trials that will most benefit them.

It is not all smooth sailing, though, and there are many skeptics. Scientists have suffered a number of agonizing setbacks in the arena of genomically targeted drugs. Compounds that caused cancer to go into remission by focusing on molecular targets in cancer cells, for example, failed to deliver a permanent cure and allowed the cancer to return in a more aggressive form. And critics of precision medicine, which is what this field is called, rightfully point out that scientists still do not understand very well how genes actually work. They point to discoveries relating to the role of so-called junk DNA, previously thought to be passive genetic material but now understood to play a significant, active role in governing biological processes.

But we will surmount these obstacles and develop new techniques and technologies, because a lot is happening at the same time. With the massive numbers of genomic data already available and the ability to sequence at will, scientists can experiment, learn from mistakes, and quickly move on to new ideas. They will decipher the complex relationships between DNA and biological processes with the help of artificial intelligence and Big Data analytical tools. Increasingly precise knowledge and deepening understanding of DNA will lead to a wholesale shift in how we think about medicine and health, and we will move from broad-stroke to personalized health care.

The Big Shift in Medicine: From Broad Stroke to Precision Genomic Targeting

In 1972, Richard Nixon declared a war on cancer. The president wanted to be able to declare cancer eradicated, much as we have since declared smallpox and polio. It was a noble but quixotic fight. Doctors understood, even then, that cancer is not a single disease, yet treatments such as chemotherapy and radiation tended to focus more on location and gross processes than on specific cellular biology.

Today, cancer remains very much with us, but investigation of this broad class of illnesses has helped spark a wholesale shift in medical thinking that extends well beyond it.

Breast cancer, for example, as not only doctors but many patients now understand, comprises a genetically diverse realm of ailments that are in many ways biologically unrelated. High-profile celebrities such as actress Angelina Jolie may opt for radical mastectomies when they discover that they carry genes that promise a high likelihood of a particularly rapacious cancer's appearance in their breasts later in life, but this may not be the most appropriate response.

Eric Green, director of the National Human Genome Research Institute, explains that cancer is essentially a genomic disease. "Instead of classifying cancers by the tissue where they are first detected—colon, breast, or brain—doctors are beginning to categorize cancer by its genomic characteristics and select treatments based on the signature of different mutations. This approach promises to treat patients with the most effective medicines while minimizing undesirable side effects, especially when chemotherapy is unlikely to help," he said to me in discussing this technology's future.

The usefulness of DNA sequencing has expanded from its role in pure research to use in diagnostics, clinical practice, and drug development broadly. The large amounts of data are enabling scientists to identify key genetic predispositions to more than 5,000 of the inherited diseases resulting from mutations in a protein-encoding gene. In an extensive research program, the Centers for Mendelian Genomics is working to find the genomic bases of these diseases, which collectively afflict 25 million Americans. Its researchers reported in a paper in August 2015 that they had identified mutations in 2,937 genes and were making as many as three discoveries a week because of "next-generation" DNA-sequencing technologies.[2]

Sloan Kettering Hospital is using IBM Watson to deliver personalized treatment plans for patients. Watson pores over all the literature, studies a myriad of drug interactions, and sifts through treatment outcomes for patients with similar genetic makeup, background, and cancer strains to identify the best possible course. This is something a physician simply cannot do in a reasonable period.

The Sloan Kettering offering highlights another key shift, to using A.I. to make the doctors smarter and let them focus on the medical aspects that require a human touch and judgment. But, though Watson may be able to tell a doctor the highest-outcome chemotherapy regime, the computer cannot help patients make a decision as to whether to continue treatment that makes them feel horrible and offers a slight chance of recovery. Medical decisions about very serious illnesses are ultimately human decisions, and for that we still very much need the help of doctors, nurses, and others with true empathy. (It will be a long, long time before A.I. can eliminate these jobs.)

The new era of precision medicine and granular understanding of the interplay of all genetic material and environmental stimuli has enlivened quests for extreme longevity. Google, for example, has launched Calico, a new company focusing on radical life extension; and Craig Venter is one of the cofounders of a company called Human Longevity, which is working on extending the healthy human lifespan through genomics-based stem-cell therapies that mitigate the diseases of aging. Venter's company is sequencing hundreds of thousands of genomes and incorporating data from functional-MRI scans that capture views of and data from processes inside a living human body in order to match genetic processes with in vivo biological ones.

The next big medical frontier after genomics is also already on the horizon: the microbiome, the bacterial population that lives inside your gut. This is a field that I am most excited about because it takes us back to looking at the human organism as a whole. Scientists are coming to the conclusion that the microbiome may be the missing link between environment, genomics, and human health. They are discovering connections between what types of bacteria live inside your body, how your genes behave, and how healthy you feel.

Microbiome: Bacterial Rainforest in Your Gut

Many children are born with genetic predispositions to type-1 diabetes. Though some of those infants become diabetic in their earlier years, others do not. A key reason for this may lie in the microbiome. In February 2015, researchers from M.I.T. and from Harvard University released the results of the most comprehensive longitudinal study yet of how the diversity and types of gut flora affect onset of this type of diabetes.[3] The scientists tracked what happened to the gut bacteria of a large number of subjects from birth to their third year in life, and found that children who became diabetic suffered a 25 percent reduction in their gut bacteria's diversity. What's more, the mix of bacteria shifted away from types known to promote health toward types known to promote inflammation.

Correlation is not causation, but the results added to evidence that the bacteria in our intestines have a strong effect on our health. In fact, manipulating the microbiome may even become more important than genomics and gene-based medicine. Unlike genomics and gene therapy, which require a relatively heroic effort to induce physiological changes, tweaking the microbiome appears to be relatively straightforward and safe: just mix up a cocktail of the appropriate bacteria, and transplant it into your gut.

One of the hottest topics in medicine has been the

successful treatment of Crohn's disease using fecal transplantation. This autoimmune disorder of the digestive tract ruins the lives of millions of sufferers in America. The solution is still being researched, and there could be other complications, but it appears to be relatively simple: take a small sample of feces from a healthy person, mix it up in a blender with some water, and give the Crohn's victim an enema of the fecal cocktail.[4] I know that it sounds disgusting; but, so far, the treatment has proven extremely effective. Similar research is being conducted on other diseases.[5]

What you eat, too, can affect what is in your gut. A study published in the journal *Nature* found that changes in diet can cause dramatic shifts in the microbiome within three or four days.[6] "We found that the bacteria that lives in people's guts are surprisingly responsive to change in diet," Lawrence David, assistant professor at the Duke Institute for Genome Sciences and Policy, and one of the study's authors, told *Scientific American*. "Within days we saw not just a variation in the abundance of different kinds of bacteria, but in the kinds of genes they were expressing." The researchers noted shifts in the volume of bile acid secreted. Most surprisingly, they found that bacteria native to food we eat can handle the bile bath and colonize our guts when we eat foods, such as cheeses or meats, that are happy homes to bacteria.

As this discovery illustrates, diet is important, but though it may help motivate them, getting people to alter their diets is exceptionally hard. (That's why most diets fail, right?) But we may be able to work around that by creating ways to change the microbiome balance in our bellies using supplements or other delivery mechanisms. And here's where a rapid pace in scientific advances pays off: a wide-ranging analysis of microbiomes to identify bacterial population patterns in the guts of millions of people could provide a blueprint for the gut contents of disease sufferers and healthy people. This census could inform treatment efforts and allow doctors to more effectively manipulate the microbiome. So a yogurt a day may keep the doctor away, and a tummy full of the right sorts of cheese may be better medicine for many metabolic syndromes than pharmaceuticals or other lifestyle changes.

Genomic medicine, gene-targeted drug development, and microbiome manipulation all depend on the ability to read DNA. But a few years ago we pushed past simply reading DNA and began writing it: creating life forms that did not naturally evolve.

Altering Life Itself: The Rise of Synthetic Biology

A computer-designed virus that cures a fatal disease, new types of bacteria capable of synthesizing an unlimited fuel supply or of wiping out every person on Earth, customized biotoxins targeting the genome of the U.S. president, engineering an Olympic athlete for height and strength from the first days of inception: science fiction? Actually, we're well down that road. When looking back in history a hundred years from now, we will probably recognize this as the greatest historical shift in genetics: the transition from reading and understanding DNA to editing it in living organisms and creating entirely new organisms using chromosomes created de novo from DNA base pairs.

In May 2010, Craig Venter announced that his team had, for the first time in history, built a synthetic life form by creating entirely novel DNA. Christened *Mycoplasma mycoides* JCVI-syn1.0, also known as "Synthia," the slow-growing, harmless bacterium was made of a synthetic genome with 1,077,947 DNA base pairs. To make Synthia, Venter's team inserted a synthetic genome into a cell containing no DNA.

The technology that Venter used to "write" the genes of this new organism is the equivalent of a laser printer that can "print" DNA. DNA has a fairly simple structure, with a double helix containing linked chains of nucleic acids. Already there are a number of DNA print providers, such as Thermo Fisher Scientific and GeneArt, that will sell DNA synthesis and assembly operations as a service. Current pricing is by the number of amino-acid base pairs—the chemical components of a gene—that are to be assembled. From 2003 to 2015, assembly costs plummeted from $4 to 20 cents per base pair; and in March 2016, one company, Gen9, offered assembly for 3 cents per base pair of long DNA constructs.[7]

It is very likely that, in the early 2030s, it will be possible to search for genetic designs on the Web, download them to your computer, and modify and adapt them to your needs. Cold- and flu-vaccine designs as well as custom cures for pandemics will be available on line globally upon release, and the process of printing them will be as easy as downloading an application on a smartphone. This technology may enable any of us to print our own treatment—or, more darkly, to become a backyard eugenicist.

You may recognize from the radio show "A Prairie Home Companion" the line "and all the children are above average." The new average will be above the old one—for those who can pay for it. Should the U.S. government subsidize eugenic improvements to ensure a level playing field when the rich have access to the best genetics that money can buy and the rest of society does not? Will we enter a time when those who can afford it live far longer, healthier lives than those of lesser means, because they can pay for a better genome?

We may face these questions sooner than you think likely, because of yet another genomics technology.

In 2014, Chinese scientists announced that they had successfully produced monkeys that had been genetically modified at the embryonic stage.[8] In April 2015, another group of researchers in China published a paper detailing the first ever effort to edit the genes of a human embryo.[9] The attempt failed, but it shocked the world: this wasn't supposed to happen so quickly. And then, in April 2016, yet another group of Chinese researchers reported that it had succeeded in modifying the genome of a human embryo in an effort to make it resistant to HIV infection.[10] The scientists used a new technique, the CRISPR-cas9 system, which was developed in the United States by Jennifer

Doudna, of UC Berkeley, and Feng Zhang, of M.I.T. Cas9 helps to snip out a piece of DNA from a cell and then enables the cell to stitch the ends back together. It can be used to edit out faulty parts of the DNA.

The rough material cost of editing a gene using CRISPR is between $50 and $100. In other words, it's a lot more expensive to go to an NBA basketball game than to edit a gene or to create a new DNA structure using CRISPR.

In the short term, scientists hope to use CRISPR to edit human genes for therapies against cystic fibrosis and other hereditary fatal conditions. In the longer term, supporters of synthetic biology point to huge potential benefits. Freed from the slow-moving confines of evolution, we could potentially edit genes and build new ones to eradicate all hereditary diseases. We might respond quickly with genetic alterations to withstand horrific epidemics such as the Spanish influenza that killed tens of millions. And we might design plants that are far more nutritious, hardy, and delicious than anything that exists today.

Considerable ethical and scientific concerns already exist over release of genetically modified organisms into the wild. Plans to release genetically crippled mosquitoes in the southern United States to reduce the risk of tropical ailments that they host, for instance, have met firestorms of concerns.[11] Altering the DNA of insects is controversial enough. The prospect of altering the genes of people—modern-day eugenics—has caused a schism in the science community. Regardless, research toward precisely that end is happening all over the world. Although we can't yet reformat our genomes as we can our hard disks, we are approaching such a capability.

Messing with our genetic material is, however, a risky thing. Leading scientists have called for a ban on editing the human genome, citing the enormous risks. The CRISPR's inventor herself, Jennifer Doudna, has come out strongly in favor of a cautious approach to modifying the human germ line. "The idea that you would affect evolution is a very profound thing," she said to the *New York Times*.[12]

Never before have I advocated slowing down technological development, but, in September 2015, I wrote a column for the *Washington Post* titled "Why there's an urgent need for a moratorium on gene editing."[13] I argued that we need to better understand the technologies and develop a consensus on what is ethical before allowing researchers to edit the DNA of human embryos.

We simply do not yet understand the potential unintended consequences of genomics editing. What if modifications result in terrible illnesses? What if they somehow modify brain chemistry to make a healthy genome of psychotic, remorseless superhumans? What if a synthetic bacterium escapes the lab and causes a widespread plague that kills millions?

A panel of distinguished scientists expressed the same concerns in December 2015. At a meeting convened by the National Academy of Sciences of the United States, the Institute of Medicine, the Chinese Academy of Sciences, and the Royal Society of London in Washington, the panel called for what would, in effect, be a moratorium on making inheritable changes to the human genome.[14] The group said that it would be "irresponsible to proceed" until the risks could be better assessed and there was "broad societal consensus about the appropriateness" of any proposed change. They left open the possibility for such work to proceed in the future by saying that as knowledge advances, the issue of making permanent changes to the human genome "should be revisited on a regular basis."

The Academies, however, have no regulatory power; what they published were just guidelines. And their recommendations were merely that gene-edited embryos should not be implanted in a woman's uterus to establish a pregnancy, supporting continuation of basic research in the area.

Beyond the ethical issues, synthetic biology may open a Pandora's box of national-security problems. Security futurist Marc Goodman says that it could lead to new forms of bioterrorism, with hitherto unseen forms of bio-toxins.[15] These bio-threats may be nearly impossible to detect, because they can be customized to the genome of a certain person or groups of people. Goodman, who has worked on cybercrime and terrorism with organizations such as Interpol and the United Nations, says that the bio-threat potential is greatly underestimated. "Bio-crime today is akin to computer crime in the early 1980s. Few initially recognized the problem, but one need only observe how the threat grew exponentially over time," he says.

If the tools are there, criminals and terrorists will exploit them. They have embraced drones and cybercrime because they are useful, and they will exploit synthetic biology. So we will need new types of defenses against hostile synthetic bio-toxins or life forms. And we will need global agreements to stop governments themselves from "engineering" the perfect athlete or soldier.

Do the Benefits Outweigh the Risks?

As you have probably guessed, I am deeply concerned about the implications of editing the human genome. Synthetic biology presents grave existential risks that we must examine and consider very closely before widely permitting artificially designed life loose outside the lab. That's because a synthetic biology experiment gone awry could unleash horrific disease or environmental damage that would be nearly impossible to stop.

Yes, I know that lives can be saved and diseases cured, and that we don't have time to waste; but there needs to be a balance. To allow these technologies to function safely outside the lab, researchers must put in place multiple mechanisms to ensure that engineered organisms can be, for lack of better words, killed on demand, and quickly. I am advocating not that we cease progress, but that we slow it down until we understand the risks—and are sure that the benefits outweigh them.

In this light, the most promising and least controversial realm of breakthroughs discussed in this chapter is the ongoing analysis of the microbiome. Since this is more about restoring ancient healthy systems in our intestines that evolved naturally, rather than about permanently or radically altering life forms, the microbiome promises to be the least risky and perhaps the most important way to affect our health and quell the very lifestyle diseases that have proven so resistant to all manner of interventions. This will benefit everyone equally, will not lead to dependence on drugs and doctors, and will be affordable by all. It is where we must focus more energy—as we develop guidelines on how to safely use gene editing. Fortunately, the U.S. government agrees about the importance of doing so: the White House launched the National Microbiome Initiative in May 2016 to foster the integrated study of microbiomes across different ecosystems and committed more than $121 million for the research.[16]

PART FOUR

Does the Technology Foster Autonomy or Dependency?

• 12 •

Your Own Private Driver: Self-Driving Cars, Trucks, and Planes

In a popular children's book called *If I Built a Car,* a fanciful fledgling engineer (who is probably about ten) waxes enthusiastically about designing a car that houses an on-board swimming pool, makes milk shakes, and can both fly and dive under water.[1] Of course, the car has a robot driver that can take over if the humans need a snooze.

We aren't getting cars that can make milk shakes or are big enough to house a decent sized swimming pool, and flying cars remain a couple of decades away. But our robot drivers are here.

There are debates in mainstream media over whether driverless cars will ever be adopted and whether we can trust our lives to a machine. A survey by the American Automobile Association in March 2016 revealed that three out of four U.S. drivers would feel "afraid" to ride in self-driving cars, and that just one in five would entrust his or her life to a driverless vehicle.[2]

When I first encountered the Google car in Mountain View, back in 2014, I had the same doubts. If I had taken the survey, I would have been in the three out of four who are afraid. And then, in July 2016, I took delivery of a new Tesla that had some of these self-driving capabilities.

At first, the thought of letting my car drive itself was indeed frightening. But the highway was almost empty, and the lanes were clearly marked, so I took the risk and engaged the autopilot function. I kept a firm grip on the wheel, because I didn't want to put my life in the hands of software. The fear lasted for just five minutes. Curiosity got the better of me, and I let go of the steering wheel to see what would happen. The car continued to drive just fine; it didn't need me. Twenty minutes later, I had one hand on the wheel and I was checking e-mail with the other as the car did the driving for me. I did take full control when the road was narrow or the terrain was uneven, but, by and large, I became as comfortable with the car's autosteer function as I am with cruise control.

Tesla's autosteer is just one step on the path to a fully autonomous car, but, as other technologies subject to Moore's Law are doing, self-driving systems are becoming exponentially more functional. All of the Teslas on the road are learning in tandem; they have collectively traveled in the millions of miles. Within three to four years, my Tesla will be driving by itself without any help from me.

This is not to say accidents won't happen. One already did, and it was a big deal; a man in Florida using Tesla's autosteer crashed into a turning truck.[3] He was killed. But compare this to the many thousands of fatal accidents that happen due to drunk driving and, at least statistically, we would all be safer if we used autosteer, even in its current less-than-perfect state.

What seems hard to do today will seem trivial in a few years, as we replace big, clunky, expensive systems with tiny, reliable, affordable autonomous software and vehicle-sensor packages. I expect that you will get used to this about as fast as I did. It will be an amazing transition, and we won't want to look back.

Few people seem to fully grasp the profound improvements in our lives that driverless cars will bring. Their adoption will slash accident and fatality rates, saving millions of lives. As well, it will remove one-third to one-half of all vehicles from city streets. A large percentage of the cars on the streets of New York, San Francisco, and London at any one time are looking for parking; but self-driving cars don't need to park: they can continuously circulate, picking up and dropping off passengers. The Earth Institute at Columbia University projects a 75 percent reduction in the cost of car ownership, because fewer shared vehicles will be necessary to provide the same service collectively that personally owned vehicles provide.[4] During peak hours, those shared vehicles will be in use 90 percent of the time. And, with no more need for steering wheels and other systems enabling human control, vehicles will be lighter and far more fuel efficient. Most important, car sharing will cost a fraction of what car ownership today costs. Owning a car for daily, personal transportation will seem impractical.

Self-driving cars will also deliver incontrovertible social benefits. With self-driving cars, the disabled will no longer struggle to find transportation; they will have an on-demand personal driver. Several years ago, as the *New York Times* in November 2014 relates it, Google's self-driving car team contacted Steve Mahan, Executive Director of the Santa Clara Valley Blind Center.[5] The team wanted feedback and let Mahan come along for test drives in earlier self-driving Prius models as well as in the latest Google car. "My experience with Google has been terrific, and I want it to happen," Mahan told the *Times*. "Everyone in the blind community wants it to happen."

Other groups will also benefit in tangible ways. Women and children will never worry about getting a cab ride late at night. Once all drivers are off the road, traffic violations will no longer be an issue, and cops will have fewer reasons to pull over cars, which should reduce instances of the currently vicious discrimination against individuals "driving while black." Teens will not face insurance discrimination as they do today, and their parents will not have to pay for the dubious privilege of teaching a teenager to drive. People living in the country will finally gain access to transportation services that put them nearly on par with their city cousins. Pedestrians will stop worrying about getting hit by cars in intersections.

Let me paint a picture of what streets will look like in an age of driverless cars. We will no longer need traffic lights: robot cars will synchronize wirelessly to time mass movements across city intersections and entries onto freeways or balletic dances around four-way stop signs. Having no human eyes behind the wheel will obviate much of the need for signaling and signage. When all the driverless cars are talking to each other, there will be no need for them to ever come to a complete halt and waste all their kinetic energy.

So we will be able to forget traffic lights—and stop signs, yield signs, lighting on freeways, and dozens of other transportation-infrastructure elements catering to human drivers. This great elimination will save many, many billions of dollars in the United States. Equally important, self-driving cars will eliminate the need to build these types of infrastructure in less developed countries in which traffic lights, freeways, and other modern traffic-control features have yet to be put in place. The future cost savings to those countries will be astronomical. In that future, the benefits of self-driving cars will be far more evenly distributed.

Reinventing the Car to Forgo a Human Driver

Eliminating human drivers will also allow automobile designers to build cars from a completely different mindset. Driverless cars will not need steering columns, brake pedals, accelerator pedals, or any of the other components drivers use for slowing or accelerating. They will not need a gearshift panel in the middle of the driver compartment or an emergency brake pedal. The A.I. system driving the car will also reduce accidents to negligible levels. Once accidents cease, there will be no heavy steel protective beams in the doors, or crumple zones. Self-driving cars will not require bumpers, seat-belt assemblages, or bulky airbags.

Dropping all of this extra mass and complexity will allow cars to be two things that we love: super efficient and super fast. Both will be important in the driverless future. Today a Tesla can drive 300-odd miles on a single charge. But eliminating its human-centric components would likely extend its charge range significantly.

Forgoing all those extra components could also make room for other things. Want to work while driving? You can put up a wide-screen thin plastic organic L.E.D. display and tap away at a full lap desk. Need a nap? Your car seat will recline all the way if there is no one sitting behind you, allowing full repose.

Another option would be to power a lot more enclosed space with the same drive-train and motor. Imagine if you could double the size of a motorhome: why own a real house, if living on the road is no longer inconvenient, uncomfortable, or cramped?

And let's consider speed. Since humans will not be driving, and A.I. will keep all the cars moving in perfect order, cars could move at ludicrous speeds, to steal a phrase from the Tesla folks.

That could greatly affect the way we live and the way we think about cities. Right now, there is a schism in the Bay Area, between San Francisco and the rest of Silicon Valley. Many companies are moving up to San Francisco because that's where their workforce wants to live. Driving from San

Francisco to Palo Alto can often take one and a half hours due to nasty traffic. Commuter train service is jam packed and also subject to delays. As a result, I see my friends in the city less and less. And the mixing and melding of ideas that comes from people's ability to move and meet suffers.

If a self-driving Tesla could move at up to 200 miles per hour on the highway and drive without stopping, a door-to-door commute from San Francisco to Palo Alto would likely be less than fifteen minutes (it's only a thirty-odd-mile distance), and the problem would be solved. Middle-class workers in New York City, priced out of the inner city, could live at the beach, in Far Rockaway, and make it to Manhattan in around ten minutes.

The economic benefits, of course, will be massive as well. We will spend less time trapped behind the wheel and more time doing creative things. Cities can free up parking spaces and parking garages for apartments. It will be like it was to get turn-by-turn GPS: once we had it, we couldn't understand how we had lived without it.

My grandchildren will ask me to tell them what it was like to drive a car in an old city. I'll tell them it was scary, dangerous, and wasteful, and that they are lucky to have a better way of living. Oh, and, by the way, those Google cars we've been talking about have yet, despite millions of miles traveled on the roads, to cause a single fatal or serious accident. The few minor accidents that did happen were because of the pesky, ill-mannered, and dangerous humans that they had to share the road with.

Moral Argument for Self-Driving Cars

Even after decades of decline, car accidents are a leading cause of preventable death in the United States. The National Highway Transportation Safety Agency surveyed tens of thousands of crashes and found that human error was a probable cause in 92.6 percent of them.[6] Worldwide, 1.25 million people died in car accidents in 2013, according to the World Health Organization.[7] In the United States in 2013, more than 32,000 people died in car crashes, despite access to one of the best emergency-healthcare systems in the world.[8] In the developing world and middle-income nations, the road-traffic death rates are twice as high as in most developed Western countries, partly because of the difficulty of getting the casualties to hospitals in the critical first hour after the accident from locations where transportation infrastructure is lacking.[9] If all of those cars outside the United States had been self-driving, in all likelihood we would have avoided 95 percent or more of those accidents and saved more than a million lives each year.

Simply put, people are poorly designed to guide two-ton hunks of metal. People drink and drive, fiddle with radios, fall asleep at the wheel, drive too fast, mistake the accelerator for the brake, and on and on and on. Many crashes are blamed on inattention. In other words, people cannot pay attention well enough to not crash. Autonomous vehicles are subject to none of these problems. The cost of self-driving systems is rapidly falling, and within a decade will be no more than $100. So there is a clear imperative to adopt self-driving cars as quickly as possible.

In the United States, big-rig trucks remain a major source of fatalities, demolishing cars in accidents. Many of these accidents happen on interstate highways, where drivers sit behind the wheel for days and frequently flout regulations on driver hours. A significant portion of these accidents, most of which are fatal to car drivers, are caused by the truck drivers falling asleep or driving with serious sleep deprivation. So it was no surprise when Daimler-Benz put the first self-driving big-rig on the road in May 2015. Approved for use in the state of Nevada, the truck will handle driving duties on highways, ceding city driving to a human driver, who remains on board at all times.

Driverless cars also provide safe options for getting home at night. Instead of having to pay for a taxi or Uber when getting home late, women (especially young women) can summon a self-driving car.

And No, It Isn't Just the United States That Will Benefit—And Lead

In this discussion, I've focused on the United States. But the benefits are even greater to the crowded and polluted cities of the developing world, which will benefit tremendously from dramatically lower energy usage, an inexpensive transportation system for all, and a reduction in traffic and smog.

The United States has no monopoly on this innovation, and China may well leap ahead. Its leading technology company, Baidu, has developed its own self-driving software. After testing its technology in Beijing and Wuhu, in China's southeastern Anhui province, Baidu obtained permission, in September 2016, from the state of California to begin testing there.[10] It won't surprise me if Baidu perfects its software before Google and Tesla do and China starts transforming entire cities into "autonomous car–only" zones.

And in August 2016, the world's first self-driving taxis were picking up passengers in Singapore. This was in a 2.5-square-mile business and residential district called one-north. Singapore being a small, landlocked island with congested roads, its transport planners are very motivated to replace its cars with robots. "We face constraints in land and manpower. We want to take advantage of self-driving technology to overcome such constraints, and in particular to introduce new mobility concepts which could bring about transformational improvements to public transport in Singapore," said Pang Kin Keong, Singapore's Permanent Secretary for Transport, to the Associated Press.[11]

A Massive Disruption Caused by Giving A.I. the Keys

Mandating autopilot for everything that moves would devastate employment in the sector. According to the American Trucking Associations, in 2010 approximately 3 million truck drivers were employed in the United States, and 6.8 million others were employed in jobs relating to trucking activity, including manufacturing trucks, servicing trucks, and other types of jobs.[12] So roughly one of every fifteen workers in the country is employed in the trucking business.

According to the U.S. Bureau of Labor Statistics, roughly another 300,000 people work as taxi drivers and chauffeurs. Those numbers would probably swell considerably if they included the new wave of part-time drivers. Uber, for example, claims over 14,000 cars in New York City alone.

For the near future, job growth in these industries is quite strong. But over time, driverless vehicles would mean the loss of close to 5 million jobs to the robots, with no obvious replacement jobs in sight.

And even though it is a near certainty that replacing human-directed vehicles with self-driving vehicles would reduce casualties, we already know that humans are more likely to blame robots when things go wrong but less likely to credit them for improvements. Researchers Tammie Kim from M.I.T. and Pamela Hinds from Stanford, examining this topic specifically, wrote in a paper titled *Who Should I Blame? Effects of Autonomy and Transparency on Attributions in Human-Robot Interaction:*

> Our results suggest that when a robot has more autonomy, people will attribute more blame to the robot and less to themselves and their co-workers. This is consistent with our prediction that autonomy will contribute to a shift in responsibility from the person to the robot. It is interesting

> to note, however, that attributions of credit did not show the same pattern. That is, people shifted blame for errors, but not credit for successes to the robot.[13]

And then there are the impacts on our cities, social structure, and industries.

When parking decks become freed up and streets become pedestrian walkways, city layout becomes more flexible. We will be able to set large parts of the cities aside for parks and recreation. With location and distance not being barriers, we will be able to live anywhere, and patterns of social interaction will change. Imagine being able to visit friends for dinner who live in a nearby city or being able to get to the beach on a weekend without any traffic delays.

The real-estate industry will surely be in turmoil as land-use patterns change and reverse urbanization happens. The industry won't be able to predict space utilization, because there are no precedents for the type of changes that will be occurring in the mid-2020s.

The automotive industry will be in decline because the number of cars purchased—even by the car-sharing companies—will fall dramatically. And then what happens to the car dealerships? When we can get from one city to another in relatively short periods in comfort in an autonomous car, why would we bother to take the train or struggle with the long security check lines at airports? For me, it's already a toss-up between driving and flying when I want to travel from San Francisco to Santa Barbara, which is four and a half hours away by car and takes four hours by plane and taxis (provided there are no flight delays). The self-driving cars will easily tip the balance; for any trips on the West coast, I'll forgo the flights. Imagine the disruptions to the railroad and airline industries when we all start making this choice.

And all of this begins to happen by the early 2020s. If I can rely on Elon Musk, my Tesla will become fully autonomous as early as 2018;[14] and Uber's CEO, Travis Kalanick, has signed a pact with Volvo to have self-driving cars on the roads by 2021.[15]

Does the Technology Foster Autonomy Rather Than Dependence?

I simply can't wait for self-driving cars to take over our roads; I see them as increasing our personal autonomy as much as, if not more than, anything else discussed in this book. Let's be honest: we may think we own our cars, but in reality our cars also own us. Buying a car is one of the most stressful processes in our lives. Fixing a car (or finding mechanics we can trust) is an equally problematic and far more common problem. Managing our auto insurance, washing our cars, maintaining our cars, and then ultimately getting rid of (selling or donating) our cars all take big chunks out of our lives. Then there is the part of our lives lost to driving a car in unpleasant circumstances, such as fighting rush-hour traffic or circling the block in a city center to look for parking.

Beyond this revisionist take on automobile autonomy, I see self-driving cars as opening up entirely new vistas. When parents can call a Google car and put their children in the back seat for a ride to soccer practice, that increases autonomy. When an elderly person who can no longer drive can call an autonomous vehicle for a lift to the supermarket or to the art museum, that increases autonomy. When all of this is affordable—so affordable that anyone can pay for it—it will have brought about a massive net increase in autonomy for all and an important increase in equity.

Yes, we will be dependent on autonomous cars, but we have always been dependent. Here, the dependency is actually replaced with something more reliable. The child always needs to get to soccer practice, whether a parent or neighbor or a Google car is providing the transportation.

As you can tell, when it comes to self-driving cars, I'm a starry-eyed optimist. The reality is that it won't be an easy journey. Look at the negative publicity that Tesla got when a driver in Florida lost his life while his car was in autopilot. He trusted the system more than he should have, and Tesla got all of the blame. There were calls to outlaw the technology. And there surely will be further fatalities because of software imperfections and human error. These will be a tiny fraction of the number of lives that use of the technology saves, but it won't matter—we will blame the machines.

We can also expect that reckless humans will cut self-driving cars off and jump lights—because they know that the cars are programmed to stop and give way. There will be street battles between man and machine.

The transition will be as traumatic as the battles between the horseless carriage (as the first cars were known) and the horses for supremacy of the roads. Of course the cars won, but we traded one set of problems and risks for another and one type of dependencies for another. And there will have to be a loss of autonomy for the drivers because we will ultimately have to take this away—they are too moody and dangerous. They will become the drivers in the driverless car.

◆ 13 ◆

When Your Scale Talks to Your Refrigerator: The Internet of Things

Your refrigerator will talk to your toothbrush, your gym shoes, your car, and your bathroom scale. They will all have a direct line to your smartphone and tell your digital doctor whether you have been eating right, exercising, brushing your teeth, or driving too fast. I have no idea what they will think of us or gossip about; but I know that many more of our electronic devices will soon be sharing information about us—with each other and with the companies that make or support them.

The Internet of Things (I.o.T.) is a fancy name for the increasing array of sensors embedded in our commonly used appliances and electronic devices, our vehicles, our homes, our offices, and our public places. Those sensors will be connected to each other via Wi-Fi, Bluetooth, or mobile-phone technology.

Using wireless chips that are getting smaller and cheaper, the sensors and tiny co-located computers will upload collected data via the Internet to central storage facilities managed by technology companies. Their software will warn you if your front door is open, if you haven't eaten enough vegetables this week, or if you have been brushing your teeth too hard on the left side of your mouth.

The I.o.T. will be everywhere, from heart-rate monitors in your watches to breathing monitors stitched into your child's pajamas. It will help us learn from our behaviors, manage our environment, and live a richer life.

But there is a really dark side to this machine vigilance. The Internet of Things will offer unprecedented spying possibilities, from the insurance company monitoring how you drive by using an accelerometer device in your car (which insurance giant Generali is already doing, under a scheme it calls *Pago como Conduzco,* "Pay as I Drive"[1]) to the little Samsung Paddle placed under your pillow that records your sleep cycles and vital signs, to the camera in your TV that gets hacked and allows people to watch you.

The possibilities for unhealthy and potentially illegal invasions of privacy grow along with the growth of the I.o.T.

The Awesome Things about the Internet of Things

The smash-hit Nest home thermostat may surprise you. What could be more boring and mundane than a thermostat, right? Yet the Nest is a beautiful glowing dial on your wall that is extremely easy to understand and adjust.

Americans waste huge amounts of money running their heaters and air conditioners when they aren't at home. Some people remember to turn these systems on and off when they are coming or going, but few are conscious of the seasonal adjustments they make to heating and cooling usage in their homes, or how they behave differently on weekends. That adds up to billions of dollars each year in wasted energy spending. It's an enormous market, one that is hard to address without building a truly intelligent and connected device.

The Nest set a new standard for smart devices. With motion sensors, the Nest monitors a user's daily movements. In the first few weeks after installation, it studies your behavior to learn your preferred home temperature. It also studies your coming and goings. You need to actively adjust the Nest during that period, but like a good soft A.I., it learns your habits and then starts to work all by itself.

By a certain point, the Nest becomes nearly 100 percent autonomous and optimizes the temperature in your home with no prompting. It reduces energy bills, perhaps by as much as 10 percent, and makes your home more comfortable.

The Nest also ties into utility programs that ask users to cut back on power usage at times when energy consumption is at a peak, to relieve pressure on the electrical grid. Nest users who live where such programs are in place can save 5 percent or more on power bills by participating. That's an early but effective instance of the smart grid, a sub-sector of the Internet of Things focusing on energy and our giant, antiquated, and inefficient power-generation and transmission system.

You can install the Nest application on your phone and control your home environment remotely. So, say, if you want to start cooling down your house fifteen minutes before you arrive home, you can send a message to the Nest. That's handy if you are coming home earlier than planned on a hot summer day in Phoenix, for example.

Since designing its thermostat, Nest, which operates as an autonomous unit inside Google, has released a smoke detector and a video camera to monitor for intruders or behavior of pets (and perhaps children or teenagers?). It will probably release many new products, all controllable from the Nest application.

Technology companies say they will use the Internet of Things in the same way: to reduce our energy usage, improve our health, make us more secure, and nudge us toward better lifestyles. Of course, the I.o.T., they say, will save us money too.

The ability to collect such data will have a profound effect on the economy. The McKinsey Global Institute, in a report titled *The Internet of Things: Mapping the Value beyond the Hype,* says that the economic impact of the Internet of Things could be $3.9 to $11.1 trillion per year by 2025: up to 11 percent of the global economy.[2]

Much of the value of the I.o.T. is hard for us to comprehend, because it will be machines talking to other machines to enable different A.I. systems to work together and make better decisions. By monitoring machines on the factory floor, the progress of ships at sea, and traffic patterns in cities, the I.o.T. will reach far beyond our homes and create value through productivity improvements, time savings, and improvements in asset utilization. The 200-mile-per-hour ride in a Google car will be controlled by a transportation subset of the Internet of Things, a web of sensors on the roadways and embedded in the vehicles that will allow them to speak the same language.

The McKinsey report also assigns value to the I.o.T. by including the economic impact of reductions in disease, accidents, and deaths. Those are real economic benefits even if they are hard to calculate today, with few of those systems in place. McKinsey believes that the I.o.T. will monitor and help manage a huge swath of activity on Earth: the natural world, people, and animals.

The Internet of Things should not only change our interactions with devices and improve their efficiencies but also create entirely new ways of understanding the global economic engine. Turning electronic products into software-controlled machines enables continuous improvements both to the machines and to the business models for using them. The constant improvement in features that we see in our smartphones will become common on our other devices.

Everything will be connected, including cars, street lighting, jet engines, medical scanners, and household appliances. Rather than throw appliances away when a new model comes out, we will just download new features. That is how Tesla is enhancing the self-driving features in its cars: learning and then sending software updates every few weeks. Through the software of the Internet of Things, everything will drive itself, upgrade itself, turn itself on and off at the right time, and know when it is about to break down.

The Frightening Thing about the Internet of Things

It was not an auspicious beginning to the 2015 holiday season. On Black Friday, we learned that a hacker had broken into the servers of Chinese toymaker VTech and lifted personal information on nearly five million parents and more than six million children.[3] The data haul included home addresses, names, birth dates, e-mail addresses, and passwords.[4] Worse still, it had photographs and chat logs of parents with their children.[5]

Earlier in the same month, Bluebox Security discovered serious vulnerabilities in Mattel's *Hello Barbie,* the Internet-connected version of the iconic doll toy.[6] The exploits raised the obvious question: as more toys become connected to the Internet, how many have lax security? And how many millions, or hundreds of millions, of children are in danger due to it? It is entirely possible that serious vulnerabilities affect the majority of Internet-connected toys.

Yes, these are the early days of hack attacks on toys, so hackers have a head start. But the bigger problem is that there is no real regulation of the Internet of Things. There are no severe fines for companies that have lax security. The companies can just get away with an apology; knowing that there is a privacy risk doesn't oblige them to do product recalls.

If you don't have children, then think about the huge amount of information about your life that Nest has access to if you install several of its products in your home. The company knows all about you, including the deeply intimate things that no one else does. Their cameras watch you 24x7. Think of what could happen if they got hacked.

I truly fear the increasing incursions on our privacy and confidentiality. It isn't just the toys and thermostats in our homes: cameras are already recording our every move in city streets, in office buildings, and in shopping malls. Our cars will know everywhere we have been, and our newly talkative devices will keep track of everything we do. Privacy will be dead, even within our homes. This is a risk that our smartphones already pose; soon we will be tracked everywhere.

And then there is the incessant marketing. Many of the new I.o.T. products and features on our devices will be inexpensive and useful—telling us when we need to order more milk, eat our medicine, rethink having that extra slice of cheesecake—but they will also tell us to order milk from Google Express and to get our prescriptions filled at CVS. Amazon's fast-growing Dash program lets users of Tide laundry detergent reorder with a push of a button on a cheap, wireless Tide I.o.T. device—but it is connected to Amazon's cloud, and orders are, of course, fulfilled via Amazon.com.

Will we be happy for the manufacturers of our refrigerators to recommend new flavors of ice cream, for our washing machines to suggest a brand of clothes to buy, for our scales to recommend new diet plans because that paleo diet just isn't working? They will have the data necessary for doing so—just as your smart TV's manufacturer is learning what shows you watch.

As well, we have no guarantees that the technology companies will not share our data with advertisers that want to hound us into buying their stuff. In fact, this is how the tech companies usually make their money—by selling our data (though, yes, they may claim to anonymize it). It's a Faustian bargain, but one that we commonly make. As they say, if a service is free on the Internet, then the customers are probably the product for sale.

I am not looking forward to having my bathroom scale tell my refrigerator not to order any more cheesecake. I will not permit security cameras to capture images from the inside of my home and upload them to the Internet, even with the best encryption known to man. But it is going to be very hard when Smart TVs and refrigerators have cameras embedded in them—as our laptops already do.

Does the Technology Foster Autonomy Rather Than Dependence?

Without doubt, with the industrial I.o.T., we will see tremendous benefit, a dependency that is good for us. That our cars tell us when they are going to break down or our traffic lights tell the town that a bulb needs to be replaced is all good. General Electric is able to remotely troubleshoot its jet engines and locomotives in order to warn operators about required maintenance; that feeds into planning software, which is a huge leap forward that will save time and money.

I also don't mind having a Nest thermostat regulate my

home's temperature. But do I need my refrigerator watching what I eat and talking to my phone? I don't think so; it will just make me a lot more dependent without providing the promised benefits. And it will compromise my privacy.

The Risks Don't Outweigh the Benefits

VTech admitted that its security had not been up to snuff and apologized after the 2015 hack.[7] But the company had little real incentive to worry about security, because there is no real bite in the laws seeking to penalize companies for failing to protect their customers' data. Even in California, where companies are legally required to quickly disclose hacks and warn customers that their data have been stolen, breaches continue. Globally, cyber attacks increased by 48 percent from 2013 to 2014, according to a large survey by consultancy PwC, and in 2014 those attacks cost each affected business, on average, $2.7 million.[8]

Even when there are voluntary recalls, it is possible that these costs are not sufficient motivators. Rarely are victims compensated for seizure of their identity, an unwanted gift that keeps on giving for many years. VTech earns $2 billion in annual revenue and says that some of its fastest areas of growth are Internet-connected children's products. A better way to deal with this might be to dramatically raise the penalties for lax security. This could be accomplished by insurance companies but should also include some mandatory payback clause to compromised customers. Or perhaps it could be a contribution system whereby all manufacturers of connected devices pay into a compensation pool. This is, of course, another flavor of insurance. Businesses would hate this idea, but it might force them to do the right thing.

Increases in government regulations are rarely productive and often harm innovation. But it may be prudent to expand the equipment-authorization program of the FCC.[9] This requires the testing of radio-frequency devices used in the United States to ensure that they operate effectively without causing harmful interference and that they meet certain technical requirements. In the future, these requirements could include the encryption of data and other security safeguards. This is particularly important given that our Internet of Things devices are mostly manufactured in China. The security holes could allow snooping on an unprecedented level—in homes as well as offices.

And I have another really radical thought, which goes beyond what I recommended in the chapter on security and privacy.

What if we mandated that businesses create systems that allow customers to control their own data—to see what is being collected and to alert them when those data are stolen? This has long been a pipe dream of privacy activists and an ideal of defenders of electronic civil society such as the Electronic Frontier Foundation. But we are actually tantalizingly close to having the capability of creating such a system.

My colleagues at Stanford Law School, along with many others, have been researching how this would work. Roland Vogl, who heads Codex, the Stanford Center for Legal

Informatics, envisages a system that will allow people to manage and analyze all of their structured data, including those generated by Internet of Things devices. End users will connect their devices to a "personal dashboard," through which they will be able to monitor and control their data. They will select which data can be shared and with which companies. Vogl says there are already some implementations of these technologies, such as OpenSensors and the Wolfram Connected Devices Project.

The solutions aren't difficult. We just need the motivation, regulation, and coordination. The alternative, in today's wild, wild west of Internet of Things development, is a runaway increase in security nightmares. It will be better to set the standards now and ensure a safer cyber world for our children and ourselves than to try locking the door once all the wrong people have our data.

This again is where you must be involved: we need the public demanding these protections. But first we must understand the key issues. You can also exercise the same choice as I am: until I am convinced that there is enough security, I am not going to be buying an I.o.T. home device.

• 14 •

The Future of Your Body Is Electric

In the television series *Star Trek,* the blind Lieutenant Commander Geordi La Forge wore futuristic goggles called a VISOR (for Visual Instrument and Sensory Organ Replacement). With the VISOR, La Forge enjoyed vision better than humans do with normal eyes.

Today, in the real world, a company called Second Sight is selling an FDA-approved artificial retinal prosthetic, the Argus II. The Argus II provides very limited but functional vision to people who have lost their sight due to retinitis pigmentosa, a retinal ailment that presently afflicts about 1.5 million people world wide. The Argus II captures images in real time with a video camera and processor mounted on eyeglasses. A wireless chip in the eyeglass rim beams the images to an ocular implant that uses sixty electrodes to stimulate remaining healthy retinal cells, and those cells then send visual information to the optic nerve. The Argus II lets people detect light and motion but not much more; users cannot recognize faces or detect colors, for example. And its cost is prohibitive, at U.S. $100,000.

On a Moore's Law curve, the time from the Argus II to La Forge's fully functional VISOR system should be about twelve years. Still, to think that hundreds of patients already walk the face of the Earth with a direct interface between a video camera and their eye is stunning. The question this raises in my mind is why, as systems like the Argus become cheaper and better, we wouldn't proactively replace our eyes with more powerful prosthetics.

The case of the Argus illustrates how the digitization of everything will have profound positive effects on our world. Vision is fragile; the miraculous biological mechanism that converts light to nerve impulses all too commonly breaks down with age or injury. For example, age-related macular degeneration (AMD) of the retina—a thin layer of light-sensing tissue on the inner eye—presently hobbles as many as two million Americans.

Addressing a broad range of ailments that affect our legs, our eyes, our hearing, and our sense of smell, the pieces are coming together to extend, replace, or improve on the human form. What will be even more shocking is the speed and low cost with which we will be able to build truly personalized, highly customized medical systems that are designed specifically to fit our bodies, our blood chemistry, our environments, and our genes.

Printing Body Parts, Saving Lives

Garrett Peterson's parents noticed from the day on which he was born that he frequently stopped breathing and turned blue. Garrett had an exceptionally weak trachea that collapsed frequently, cutting off his air. This is a rare condition, and one that can be fatal. Any stress, including diaper changes or crying, could result in Garrett's near asphyxiation.[1]

Garrett was sixteen months old and on the verge of dying from lung damage. His parents approached Scott Hollister, a biomedical engineer at the University of Michigan who had designed many plastic implants. Hollister designed a plastic splint to fit perfectly into Garrett's throat and hold his windpipe open. He printed the implant with a 3-D printer. The surgery was a success. The splint held Garrett's trachea open and saved his life. No longer starved for air, he quickly strengthened. Hollister designed the splint to expand as Garrett grows and to eventually dissolve, when the trachea is strong enough to support itself. "He's being more interactive and more alert and reaching more for his toys. He's just starting to be more like a normal child," Garrett's father, Jake Peterson, told National Public Radio in March 2014.[2] Rather than wait for months or years for a medical-device company to build an implant and run it through the approval process, the Petersons and Hollister built it themselves, and the design and 3-D-printing process took a week and cost less than $10,000. This touching story is only one of many describing how do-it-yourself 3-D printing and other do-it-faster or -cheaper techniques are transforming previously costly tasks into quick, easily replicable exercises. One can readily envision a company emerging in China or the United States that specializes in experimental artificial implants machined with 3-D printers, charging only a few hundred dollars. The implants could be delivered overnight. Or you might print them on your own 3-D printer.

This pairing of personalized 3-D printing with medical prosthetics or enhancement is happening in many places. For example, Ekso Bionics designs and makes robotic exoskeletons to help the paralyzed walk again. Many of Ekso's customers now pair the exoskeleton with 3-D-printed interface parts to make using the computerized, robotic legs easier and more comfortable.

The new legs have not only freed some Ekso users from wheelchairs but also taken them to places that would have seemed unreachable using older generations of technology. On February 27, 1992, Amanda Boxtel had a freak skiing accident in Colorado.[3] An expert skier, she crossed the tips of her skis on an intermediate slope and flipped over, landing on her back on the frozen ground. The spill shattered four vertebrae, leaving her legs paralyzed. In 2012, she received access to an early prototype of Ekso Bionics's exoskeleton. The first time she stood up and walked, tentatively, heel to toe across the room, she wept. And she knew that in her lifetime she would be able to walk on her favorite beach, hike in the mountains, and maybe, just maybe, even ski again. The computers that coordinate her legs are getting smarter and better; the exoskeletons, stronger, lighter, and less intrusive: software and hardware advancing together at an accelerating rate, like a skier gathering speed for a free and easy downhill run.

The $6,000 Man (and Woman): The Inevitable Progress from Human to Superhuman

The 1970s television series *The Six-Million Dollar Man* was a favorite of mine. Its lead character, Steve Austin, was an astronaut who had suffered a catastrophic accident. The government experimentally rebuilt him, giving him legs with which he could run at sixty miles per hour, a telescopic eye that could magnify human vision by twenty times, and an arm with the strength of a bulldozer. The procedure cost the U.S. government $6 million. That's probably close to $350 million today, adjusting for inflation. But what if, in a decade and a half, we could build him for $6,000? That is what I believe it will cost (in '70s dollars).

The body parts need not be made of plastic; 3-D printers can already print out biological materials. Called bioprinting, the process uses so-called bio-ink, a term that describes multi-cellular building blocks and materials that can serve as a scaffolding.

The tissue designer suits the scaffolding design to the target tissue, using inert gels to support fragile cell structures or to create gaps, channels, or void spaces that reproduce physical features of natural tissue, and develops the bioprocess protocols to create from the appropriate cells a bio-ink from which the body will make the target tissue. After designing and testing the bio-ink mixture, the designer can load up the bioprinter and, layer by layer, build the desired tissue structure.

In a paper released in April 2014, Wake Forest University scientists and regenerative-medicine expert Anthony Atala described several successful trials to create de novo vaginas for young girls whose vaginas had been missing or malformed at birth.[4] According to the trial results, the implanted vaginas, built using 3-D printing and bio-ink, worked quite well.

Atala has also bio-printed bladders in clinical trials, and is working toward creating more-complex organs with detailed vascular structures, such as kidneys and livers. Adding the blood vessels necessary to feed the tissues or organs that metabolize or process nutrients and toxins remains elusive; that could be decades away. But, within the next decade, any relatively simple structure that you might need replaced—heart valve, bone, ear, or nose—will likely be grown using a mixture of bio-ink and, to minimize risk of rejection, your own stem cells. There are now nearly a dozen companies working toward bioprinting real body materials; and industrial design giants such as Autodesk, led by its visionary CEO, Carl Bass, are researching the fabrication of biological materials through additive printing.

Of course, we will need doctors and surgeons to do all this; it is not like printing and installing a spare part in your car. But these are the technologies that our medical practitioners will commonly use.

As we gain the ability to grow tissues and print organs, we also gain an unprecedented ability to create hybrid materials and weave together biology and chemistry. Nanomaterials are of particular interest in this regard; for example, researchers are studying bone regeneration with nanostructured calcium phosphate biomaterials. The calcium phosphate acts as scaffolding and mimics crystallographic properties of inorganic components of bone. Early findings have shown that these nanostructured materials, when combined with stem cells, can accelerate bone regeneration.

Alongside better-than-human materials will come sensor systems with miniature electronics that turn our bodies into minutely measured machinery. In August 2013, the United States Patent and Trademark Office issued Endotronix a patent for its wireless sensor reader for continuous monitoring of pulmonary-artery pressure. The tiny implanted biosensor is delivered to the artery via routine, minimally invasive, low-cost catheterization. The implant requires no batteries and does not necessitate puncturing patients' vessels in order to place leads. Instead, patients can hold a smartphone up to their chests, take the measurements, and send the results off to their doctors. Till now, measuring pulmonary-artery pressure—a key indicator of congestive heart failure—on a continuous basis has been difficult and highly invasive.

The movie *Fantastic Voyage* and its novelization by Isaac Asimov, in 1966, painted a picture of the promise of wireless sensors floating in our bodies and beaming back data. The Endotronix system is one of the very first to start delivering on this promise, and it will be viewed as crude by future researchers. They are already building tiny sensor systems that will allow for true in vivo measurement of many key biological functions. In effect, we will be wired up with an early-warning system to continuously assess our physical condition, preempt serious health problems, and enable prescription of proactive behaviors and therapies. We will also finally gain deep insights into the validity of many deeply held but unproven assumptions underpinning the systems theory that doctors use to describe our bodies, such as the effect of systemic inflammation on our overall health.

As we wire up our bodies, biohacking becomes a branch of computer hacking, and the sensors that help keep us alive may become a means by which to kill us. Concerned security researchers have already demonstrated that it is possible to hack into the circuitry of pacemakers. As our bodies and their electronic extensions and implants become more intertwined, cyber attacks both on our physical beings and on the information that these sensors generate will become a genuine concern.

Some will ask why we would replace parts of our body with our own cells when, in many cases, a better, longer-lasting, more durable digital alternative exists. Why replace our nose with the same old one, when we could replace it with one that is resistant to sunburn and maybe has additional nerve cells that can better detect smell and enhance our sense of taste? Why simply replace a broken bone with a replica, when we could replace it with a strong bone laced with graphene and bearing wireless sensors to track the healing process along with micro-capsules of anti-inflammatories to reduce swelling around the joint? Why replace a cornea with another one, when we could replace it with a high-resolution camera offering us perfect sight? Why just be human, when we could be superhuman? We are going to be faced with some very interesting choices.

Even as Ekso Bionics, the company that produced Amanda Boxtel's exoskeleton, continues to improve its product, dozens of companies in different disciplines are attacking problems that, if addressed, could allow Ekso to slash the price of its system and shrink it in size. New robotics components could make it much faster to build Ekso-like systems and shrink to iPhone-size the clunky backpack of wires and silicon that Amanda carries.

Speculation about digital enhancement has extended even into the realm of the brain. Recent articles have called into question the value of rote learning, asking whether we should bother memorizing material that we can easily look up via Google. And some of the staunchest advocates for transhumanism often cite Google and search technology as an extension of our brains, a way to make it far easier for us to store, parse, and recall information as needed.

To date, there has always been a relatively clear line between what was human and what was technology. Even sophisticated technology, such as cochlear implants and pacemakers, sought to work with our existing tools, not necessarily to supplant them with a more powerful version.

In the late 2020s, we will have not only crossed that line but sprinted past it.

If we become entirely dependent on these electronic crutches, will they degrade our inherent evolutionarily hard-won capabilities so far that we could no longer recoup them should the figurative plug be pulled on our extended being?

On the other hand, as we learn to create digital proxies for biological and biophysical processes, could our ability to permanently cure or mitigate the most debilitating ailments give us the digital keys to something akin to the fountain of youth?

Does the Technology Foster Autonomy Rather Than Dependence?

For paraplegics or people who would otherwise die or become blind without the help of innovative new technology, clearly the new "body electric" represents a tremendous leap forward in autonomy without any measureable accompanying increase in dependency. In these cases, the body electric can replace the broken and the dysfunctional with something that is better and that may eventually be even better than the original human editions.

Where we start to cross into territory that is less obvious is in incorporating technologies for enhancement that is more cosmetic or merely desirable rather than necessary. This can be particularly troubling if the artificial component replaces the human component—a digital eye for an analog eye, for example. This would be a one-way journey, for the most part: a one-way journey that might increase autonomy in some ways but, by and large, will make us most intimately dependent on our devices. That dependency should be carefully considered, because converting our sensory organs and limbs and muscles and even our brains to digital extensions, while incredibly enticing, opens us up to many new risks (some discussed in earlier chapters). For digital devices, there are bits and bytes, software, and many, many failure states. Sudden failure, we should remember, could be catastrophic.

No doubt, some athletes will do this, and the doping scandals of today will become the bionics scandals of tomorrow. Some people will want the enhancements that Steve Austin had so they can climb mountains and observe nature more closely. Whether this is good or bad is what we will be debating in the late 2020s and 2030s.

And, of course, on the technologies just emerging, there are no long-term studies that tell us what will happen many years in the future. We are taking big risks. And we are becoming entirely dependent on electronic crutches that could degrade our inherent evolutionary capabilities so far that we couldn't recover them.

So we will be making trade-offs and upgrading ourselves with technologies that make us less and less human. But is that really different than the medical choices we make today? When we get eyeglasses, we commit ourselves to a lifetime of medical enhancement. When I had my heart attack in 2002, I chose to have surgery and had drug-eluting stents placed in my arteries. These slowly release a drug to block cell proliferation. My cardiologist warned me that this was new technology and there were no very-long-term studies of their efficacy yet. He said that he believed that these stents were substantially better than the older bare-metal stents, but that I would have to take blood thinners to prevent sudden stent closure due to clotting—perhaps for the rest of my life. I chose the risk and accepted the fact that I would be taking a regimen of heart medications for as long as I lived. It was the trade-off that has allowed me to live a healthy and more-or-less normal life.

These are the types of choices, for better health and for longer lives, that we will all be facing.

15

Almost Free Energy and Food

Ever since the oil crisis of October 1973, when the members of the Organization of the Petroleum Exporting Countries proclaimed an embargo and caused the price of oil to increase from $3 to $12 per barrel, the world has been in a constant state of fear of impending shortages of energy and consequent price hikes. We have begun to believe that the planet will soon run out of oil and will therefore be out of energy. Governments have been jockeying to secure oil shipments. In order to preserve the Earth's dwindling energy supplies, the United States has mandated increases in the fuel efficiency of cars.

Certainly the Earth's stock of burnable fossil fuels is limited. But we have come to apply the same scarcity thinking to water; even experts believe that large parts of the planet will run out of water and that wars will break out over access to the limited supplies. Despite a wet 2015 El Niño year, the California drought is causing a fear that agriculture will have to be permanently curtailed, leading to long-term shortages of fruits, vegetables, and nuts.

Access to clean water is one of the most serious problems in the developing world. According to the World Health Organization, 1.8 million people die every year from diarrheal diseases.[1] Of these victims, 90 percent are children under five, mostly in developing countries. Eighty-eight percent of these cases are attributed to unsafe water supply and sanitation.

It's not shortage of water per se that is the problem; it's access to clean water. Water obtained from rivers and wells is infested with deadly bacteria, viruses, and larger parasites. These could be killed by simply boiling the water, but the energy necessary to do that is prohibitively expensive, so people die or suffer.

Yet we consume energy at an average rate of only eighteen terawatts, a miniscule fraction of the 174,000 terawatts of power available.[2] And that is readily supplemented by wind, geothermal, and tidal energy. The Earth is literally bathed in energy. It's the same with water; with 71 percent of its surface covered by water, Earth is a water planet. An extraterrestrial watching our news reports would think that humans are either crazy or stupid.

The problem in supply of both energy and water, so far, has been in the economics of our ability to harness solar energy—which is practically everywhere. If we could capture enough of it, we would never have to worry about shortages of energy, water, or food. We could boil as much water as we needed from the oceans, sanitize the water that is on land, and grow unlimited supplies of food.

All of this is becoming possible.

Take the problem of water sanitization. My own son Tarun has been working with a Chilean inventor, Alfredo

Zolezzi, to commercialize a plasma-based water purification technology (PWSS) that Zolezzi's company, AIC Chile, has developed. It could give households in the developing world a source of nearly endless potable water. The marvel of the technology is its simplicity. It works by applying pressure and electricity to a continuous stream of water and converting it into a plasma state. In a matter of seconds, the system can eliminate 100 percent of bacteria and viruses, which are responsible for the majority of water-related illness. It uses less energy than a hair dryer does and could be powered by a solar panel.

Instead of retrofitting an industrial solution for the needs of the poor, Zolezzi specifically created PWSS for the slums of the world—and it worked. Hundreds of children in Chile have a reliable source of clean drinking water for the first time. Mexico's leading provider of individual and integrated water technologies, Grupo Rotoplas, made a sizeable investment in AIC Chile in mid-2016,[3] took over its operations, and plans to bring the technology to the whole of Latin America. It has reduced the cost of a "camp unit," which can provide enough water for a small village, to $6,000 and expects to get it much lower. That is affordable, and I am very hopeful that it will now reach Africa and Asia—and save tens of millions of lives. But all of this depends on the advance of solar production so that there is energy to power it.

What blocked our ability to tap the sun until recently was the cost of capturing its energy and converting it into electricity (and, ultimately, heat). But a few things have changed since the 1980s. We have become much better at making semiconductors for computers; and those same pieces of silicon are what convert solar energy into electricity. We have developed ways to make solar panels from thinner slivers of silicon. We have gotten much better also at figuring out how to squeeze more out of the solar energy we capture. And, most important, economies of scale are beginning to affect the price. As more solar panels are installed, more are manufactured, and panel- and component-manufacture costs keep falling.

For these reasons, solar-energy capture is advancing on an exponential curve. With that advance, we are heading into an era of practically unlimited, clean, almost free energy. Ramez Naam explains the trend very well in his book *The Infinite Resource: The Power of Ideas on a Finite Planet:*

> When Ronald Reagan took office in 1980, average retail electricity costs in the United States were around 5 cents a kilowatt hour (in today's dollars). Electricity produced from wind power, on the other hand, cost around ten times more, at 50 cents a kilowatt hour. And electricity from solar power cost 30 times more, at around $1.50 per kilowatt hour.
>
> How the times have changed. Today, new wind power installations in good locations are producing electricity at an unsubsidized cost of 4 cents per kilowatt hour, lower than the 7 cents per kilowatt hour wholesale prices of new coal and natural gas electricity. Solar has dropped as much and is still dropping. Large-scale solar installations in the very sunniest

areas are down to 6 cents per kilowatt hour without subsidies, and are still dropping.[4]

Those figures are constantly changing. As of August 2016, Naam tells me, the unsubsidized cost of solar capture in the sunniest areas in the United States is just 4 cents per kilowatt hour (kWh), and that of wind capture is even less. In September 2016, Abu Dhabi received a bid for 2.42 cents per kWh.[5]

The first solar photovoltaic panel built by Bell Labs in 1954 cost $1,000 per watt of power it could produce.[6] Today, solar modules cost about 50 cents per watt. According to what is known as Swanson's Law, the price of solar photovoltaic modules tends to fall by 20 percent for every doubling of cumulative shipped volume. The full price of solar electricity (including land, labor to deploy the solar panels, and other equipment required) falls by about 15 percent with every doubling.

The amount of solar-generated power has been doubling every two years for the past forty years—as costs have been falling.[7] At this rate, solar power is only six doublings—or less than fourteen years—away from being able to meet 100 percent of today's energy needs. Power usage will keep increasing, so this is a moving target. Taking that into account, inexpensive renewable sources can potentially provide more power than the world needs in less than twenty years. This is happening because of the momentum that solar has already gained and the constant refinements to the underlying technologies, which are advancing on exponential curves. What Ray Kurzweil said about Craig Venter's progress when he had just sequenced 1 percent of the human genome—that Venter was actually halfway to 100 percent because on an exponential curve, the time required to get from 0.01 percent to 1 percent is equal to the time required to get from 1 percent to 100 percent—applies to solar capture too.

It isn't just solar production that is advancing at a rapid rate; there are also technologies to harness wind, biomass, thermal, tidal, and waste-breakdown energy, and research projects all over the world are working on improving their efficiency and effectiveness. Wind power's price is now competitive with the cost of energy from new coal-burning power plants in the United States, according to Bloomberg New Energy Finance.[8]

Critics of clean energy, especially those from the oil industry, argue vehemently that the sun doesn't shine at night and winds don't blow twenty-four hours a day. They say that the Achilles heel of these technologies is the ability to store energy, because batteries are prohibitively expensive and big and bulky.

The critics are wrong on this front as well, because the cost of energy storage is also plummeting. Since 1990, the cost of batteries has fallen by a factor of roughly twenty. On current trends, the price of batteries and other energy-storage techniques will fall to just a few cents per kWh by the time solar and wind have matured, making energy from the sun and wind available 24/7 and cheaper than electricity from any other source.

The advances are exceeding expectations. In a study published in *Nature Climate Change*, Bjorn Nykvist and Mans Nilsson, of the Stockholm Environment Institute, documented that, from 2007 to 2011, average battery costs for battery-powered electric vehicles fell by about 14 percent a year.[9] This decline put battery costs in 2016 right around the level that the International Energy Agency predicted they would reach in 2020. Electric vehicles will soon cost substantially less to operate, from cradle to grave, than gasoline-fueled ones. And the same technology that is used for car batteries can be used for homes and businesses to store solar energy.

Tesla is taking the lead in developing battery technologies. In July 2016, it opened its $5 billion Gigafactory, which will produce 35 gigawatt-hours (GWh) of battery storage a year—exceeding the capacity of all the lithium-ion batteries produced world wide in 2013. At its launch, Elon Musk said that he's confident the batteries will reach a price of $100 per kWh by 2020 (the average price was $1,200/kWh in 2010). Tesla is also building a version of its battery technology for use in home and business, the Powerwall, which will allow homes that capture solar energy to be completely off the grid—not dependent on the utility company even to store energy.

By the way, many new solar technologies are in development. For example, scientists are experimenting with a new material called perovskite, a light-sensitive crystal that has the potential to be more efficient, less expensive, and more versatile than any solar solutions to date. Over the past five years, perovskite's conversion efficiency has increased from 4 percent to nearly 20 percent, making it the fastest-developing technology in the history of photovoltaics. In comparison with silicon's theoretical limit of about 32 percent, the theoretical limit of perovskite's conversion efficiency is estimated to be about 66 percent, so it could be transformative if commercialized successfully.

How This Benefits Everyone, Everywhere

The effect of these advances is not limited to the developed world; it is anywhere where people can put a solar panel on a roof. Free power will trickle down even to remote villages, with profound consequences. This is already happening.

In Africa, 1.2 billion people have no connection to a power grid, and another 2.5 billion can get power only intermittently. To make matters worse, the lack of viable electrical options creates perverse side effects. People use kerosene for lamps, a dirty fuel that, according to the *Economist*, costs $10 per kWh of energy that it provides—significantly more costly than the same unit of power in the West on a modern power grid.[10] Worse, kerosene fires are endemic in Africa, and their toxic fumes cause respiratory ailments that kill hundreds of thousands per year.

The plummeting cost of photovoltaic panels, along with the decline in the prices of light-emitting diodes (another semiconductor product), has brought light to more than 20 million Africans in the past decade. The World Bank's Lighting Africa program is doubling sales of approved devices each year.[11] Solar-powered LED lamps with included battery storage sell for $8.[12] That's still a lot of money for the poorest to afford, but it's within reach.

Central power grids will probably never be built to cover all of Africa. Power there will truly be a distributed endeavor. Schools, hospitals, and homes will all be powered by sources on site or nearby. The same happened with landline communications: Africa leapfrogged into cell-phone networks. In some places, these networks are better than those in the United States. By leapfrogging legacy infrastructure and focusing on the future, Africa will be able to take far better advantage of future price declines in solar, LED, and other energy-capture and -saving technologies.

Aside from its effect on lighting, distributed micro-generation in Africa will also allow cheaper charging of cell phones. This is, believe it or not, a major expense for many Africans who lack sources of electrical energy: they pay dearly for electricity at kiosks. By reducing the cost of phone ownership and making voice and data communication cheaper, low-cost electricity boosts a key service that lifts people out of poverty and improves their lives. Information is power: to get the information, you need the power. Within a decade, we should see 50 percent penetration of solar panels into Africa and total penetration of LEDs or close access to cheap electricity for running small household appliances or charging phones.

So everywhere on Earth, for rich nations and poor nations, there will be light for all, and it will be essentially free. This will lead to many other benefits.

Free Power Means a More Peaceful Planet

Water and energy are the natural resources at the heart of many of the worst global conflicts. In the Ukraine, a core part of the dispute with Russia is over natural-gas pipelines. Japan started World War II in part due to its lack of natural resources, among them oil. India and China are tussling over water rights, a dispute that looks set to radically worsen as China seeks to expand agriculture in its south and India also pushes to grow enough food to satisfy its fast-growing population. China is proposing massive dams on major rivers flowing from China to India and Bangladesh.[13]

With cheaper power making water more abundant, even more of the desert may blossom in green edibles. The world has plenty of desert with plenty of natural sunshine for farms. Israel has pioneered desert agriculture, and tomato farms in Arizona are some of the most productive in the world. Adding water to these vast deserts, far cheaper than fertile fields, will allow many arid countries to become efficient producers of crops. Vertical farming also has great potential. Imagine turning those city parking lots that are no longer needed because of self-driving cars into farms that grow organic food with LED lights and artificial-intelligence software—organic because when food is grown in buildings surrounded by glass, we have no use for insecticides or pest control.

In his book *Abundance: The Future Is Better Than You Think,* Peter Diamandis wrote about an era in which all the needs of humanity are met: a world in which no one on Earth suffers from hunger or lacks clean water; a world in which we all have clothes, electricity, cell phones, and housing; and he believes that this is an eminently achievable aim.[14] I agree with him—if we do things right, if we can find a way of sharing the benefits of technology advances, and if we take the right paths.

I know I am making many leaps of faith in this chapter with the assumption that technology will fix all and that we will be able to get it to the right places. But when I look at how countries such as India and Africa are being transformed by cell phones, Internet access, solar energy, and education, I see the possibilities. When I spend time with entrepreneurs who are building these technologies, I see a determination to solve any problems confronting them—because it is about uplifting humanity. When I see the progress we are making with electric cars, solar energy, and battery storage, I become convinced that in the 2020s most of us in the prosperous world will have the choice of living in the same clean-energy future as I already live in—with the sun providing 100 percent of our energy needs.

Does the Technology Foster Autonomy Rather Than Dependence?

Nearly free energy and water will be, along with self-driving vehicles, the biggest boosts to autonomy that humans have enjoyed in history. Energy and water are the key to everything that offers us a more comfortable life. Energy keeps us warm, powers our vehicles, lights our homes, powers our communications systems, and much more. Inexpensive energy will also unlock an endless supply of fresh water and allow us to grow more food.

Combined, energy and water will give us as much as we could ever want or need. In those parts of the world that are poorly governed or have poor infrastructure, inexpensive energy and water will also allow people to experience lives of a quality far closer to that of us in the West and the developed world. There is no autonomy tradeoff; almost free energy and water will give us more autonomy and reduce our dependency. More than anything else discussed in this book, the ease of accessing energy and water will deliver a base level of abundance that will improve the well-being of all people on the planet, from the richest to the poorest.

CONCLUSION

So Will It Be *Star Trek* or *Mad Max*?

If, after reading this book, you complain that I have taken you on a roller coaster ride, getting you really excited about the amazing future in one paragraph and then scaring the crap out of you in the next, I will not be surprised. That is the path that technology is on: with amazing possibilities to uplift mankind, yet with really dark downsides too. There is no clear outcome: the future hasn't happened yet; it will be what we make it. As I have been arguing, it is the choices that we all make that will determine whether we build the utopian *Star Trek* future or end up in the *Max Max* dystopia.

The oldest technology of all is probably fire, even older than the stone tools that our ancestors invented. It could cook meat and provide warmth; and it could burn down forests. Every technology since this has had the same bright and dark sides. Technology is a tool; it is how we use it that makes it good or bad.

I showed you a broad range of technologies and asked you to view them through a lens or filter to assess their value to society and mankind. I asked you to consider whether they had the potential to benefit everyone equally, what the risks and the rewards were, and whether the technology more strongly promotes autonomy or dependence. It is fairness and equality that are at the heart of these questions. Industry disruption is going to happen; tens of millions of jobs are going to disappear; our lives will change for the better and for the worse. If we manage it equitably and ease the transition and pain for the people who are most affected and least prepared, we can get to *Star Trek*: we can be living in an era in which every one of us has food, shelter, education, and light, and is connected to all. We will have better lives if we can adapt quickly enough—psychologically, socially, ethically, and legally—and adjust to a world that changes literally before our eyes, every day, every minute.

At the end of the day, I believe that you will figure this out and we, as a collective race, will figure this out. Despite my fears, I know that humanity will rise to the occasion and uplift itself because it always has. We wouldn't have gotten this far if we did not have it in us to rise to great occasions.

Of the *Star Trek* future, Captain Picard once said: "The acquisition of wealth is no longer the driving force in our lives. We work to better ourselves and the rest of humanity." That is the future that we must build together.

NOTES

INTRODUCTION

1. William Gibson, speaking at interview, "Talk of the Nation," National Public Radio 30 November 1999, http://www.npr.org/programs/talk-of-the-nation/1999/11/30/12966633, Timecode 11:55. (accessed 9 December 2016)

PART ONE

CHAPTER ONE

1. "SOPA/PIPA: Internet Blacklist Legislation," Electronic Frontier Foundation (undated), https://www.eff.org/issues/coica-internet-censorship-and-copyright-bill (accessed 21 October 2016).

2. "H.R. 3261—Stop Online Piracy Act," U.S. Congress, https://www.congess.gov/bill/112th-congress/house-bill/3261 (accessed 21 October 2016).

3. "S. 968—PROTECT IP Act of 2011," U.S. Congress, https://www.congress.gov/bill/112th-congress/senate-bill/968 (accessed 21 October 2016).

CHAPTER TWO

1. James Cook, "London taxi company Addison Lee is battling to stay relevant in the age of Uber," *Business Insider Australia* 18 December 2015, http://www.businessinsider.com/addison-lee-cto-peter-ingram-explains-how-its-technology-works-2015-12 (accessed 21 October 2016). Jim Edwards, "Addison Lee's CEO told us how Uber is hurting his business and what he's doing about it," http://www.businessinsider.com.au/liam-griffin-ceo-of-addison-lee-on-how-uber-has-hurt-his-mini-cab-business-2015-4 (accessed 21 October 2016).

2. Ben Marlow, "Addison Lee owner flags sale," the *Telegraph* (U.K.), 28 June 2014, http://www.telegraph.co.uk/finance/newsbysector/banksandfinance/10933273/Addison-Lee-owner-flags-sale.html (accessed 21 October 2016).

3. Johana Bhuiyan, "Why Uber has to be first to market with self-driving cars," *Recode* 29 September 2016, http://www.recode.net/2016/9/29/12946994/why-uber-has-to-be-first-to-market-with-self-driving-cars (accessed 21 October 2016).

4. Alison Griswold, "Uber wants to replace its drivers with robots. So much for that 'new economy' it was building," *Slate* 2 February 2015, http://www.slate.com/blogs/moneybox/2015/02/02/uber_self_driving_cars_autonomous_taxis_aren_t_so_good_for_contractors_in.html (accessed 21 October 2016).

5. Ray Kurzweil, *How to Create a Mind: The Secret of Human Thought Revealed,* New York: Viking, 2012.

6. Ray Kurzweil, "The law of accelerating returns," *Kurzweil Accelerating Intelligence* 7 March 2001, http://www.kurzweilai.net/the-law-of-accelerating-returns (accessed 21 October 2016).

7. Dominic Basulto, "Why Ray Kurzweil's predictions are right 86% of the time," Big Think 2012, http://bigthink.com/endless-innovation/why-ray-kurzweils-predictions-are-right-86-of-the-time (accessed 21 October 2016).

8. Tom Standage, "Why does Kenya lead the world in mobile money?" the *Economist* 27 May 2013, http://www.economist.com/blogs/economist-explains/2013/05/economist-explains-18 (accessed 21 October 2016).

9. Peter Diamandis and Steven Kotler, *Abundance: The Future Is Better Than You Think,* New York: Free Press, 2012, p. 9.

CHAPTER FOUR

1. Tim Kise, "Uber: Congress' [*sic*] new private driver," Hamilton Place Strategies 11 November 2014, http://hamiltonplacestrategies.com/news/uber-congress-new-private-driver (accessed 21 October 2016).

2. Alberto Gutierrez, "Warning letter," U.S. Food & Drug Administration 22 November 2013, http://www.fda.gov/ICECI/EnforcementActions/WarningLetters/2013/ucm376296.htm (accessed 21 October 2016).

3. "Uber banned in Germany as police swoop in other countries," BBC News 20 March 2015, http://www.bbc.com/news/technology-31942997 (accessed 21 October 2016).

4. Personal communication with author.

5. James A. Dewar, *The Information Age and the Printing Press: Looking Backward to See Ahead*, Santa Monica, California: RAND Corporation, 1998, http://www.rand.org/pubs/papers/P8014.html (accessed 21 October 2016).

PART TWO

CHAPTER FIVE

1. Gustavo Diaz-Jerez, "Composing with melomics: Delving into the computational world for musical inspiration," *LMJ* December 2011; 21:13–14, http://www.mitpressjournals.org/doi/abs/10.1162/LMJ_a_00053 (accessed 21 October 2016).

2. Ian Steadman, "IBM's Watson is better at diagnosing cancer than human doctors," *WIRED* 11 February 2013, http://www.wired.co.uk/article/ibm-watson-medical-doctor (accessed 21 October 2016).

3. Vinod Khosla, "Technology will replace 80% of what doctors do," *Fortune* 4 December 2012, http://fortune.com/2012/12/04/technology-will-replace-80-of-what-doctors-do (accessed 21 October 2016).

4. Daniela Hernandez, "Artificial intelligence is now telling doctors how to treat you," *WIRED* 6 February 2014, https://www.wired.com/2014/06/ai-healthcare (accessed 21 October 2016).

5. Thomas H. Davenport, "Let's automate all the lawyers," *Wall Street Journal* 25 March 2015, http://blogs.wsj.com/cio/2015/03/25/lets-automate-all-the-lawyers (accessed 21 October 2016).

6. Kevin Kelly, "The three breakthroughs that have finally unleashed AI on the world," *WIRED* 27 October 2014, http://www.wired.com/2014/10/future-of-artificial-intelligence (accessed 21 October 2016).

7. Matt McFarland, "Elon Musk: 'With artificial intelligence, we are summoning the demon,'" *Washington Post* 24 October 2014, https://www.washingtonpost.com/news/innovations/wp/2014/10/24/elon-musk-with-artificial-intelligence-we-are-summoning-the-demon (accessed 21 October 2016).

8. Rory Cellan-Jones, "Stephen Hawking warns artificial

intelligence could end mankind," BBC 2 December 2014, http://www.bbc.com/news/technology-30290540 (accessed 21 October 2016).

9. "Hi Reddit, I'm Bill Gates and I'm back for my third AMA. Ask me anything," *Reddit*, https://www.reddit.com/r/IAmA/comments/2tzjp7/hi_reddit_im_bill_gates_and_im_back_for_my_third (accessed 21 October 2016).

10. The White House, "The Administration's Report on the Future of Artificial Intelligence," The White House 12 October 2016, https://www.whitehouse.gov/blog/2016/10/12/administrations-report-future-artificial-intelligence (accessed 21 October 2016).

11. Executive Office of the President National Science and Technology Council Committee on Technology, *Preparing for the Future of Artificial Intelligence*, Washington, DC: The White House, 2016, https://www.whitehouse.gov/sites/default/files/whitehouse_files/microsites/ostp/NSTC/preparing_for_the_future_of_ai.pdf (accessed 21 October 2016).

CHAPTER SIX

1. Sugata Mitra, "Kids can teach themselves," TED February 2007, http://www.ted.com/talks/sugata_mitra_shows_how_kids_teach_themselves (accessed 21 October 2016).

CHAPTER SEVEN

1. Richard Dobbs and James Manyika, "The obesity crisis," *The Cairo Review of Global Affairs* 5 July 2015, https://www.thecairoreview.com/essays/the-obesity-crisis (accessed 21 October 2016).

2. "Density of physicians (total number per 1000 population, latest available year)," WHO (undated), http://www.who.int/gho/health_workforce/physicians_density/en/ (accessed 21 October 2016).

3. Andis Robeznieks, "U.S. has highest maternal death rate among developed countries," *Modern Healthcare* 6 May 2015, http://www.modernhealthcare.com/article/20150506/NEWS/150509941 (accessed 21 October 2016).

4. Emily Cegielski, "In parts of the US, maternal death rates are on par with sub-Saharan Africa," *New York Times* 24 April 2015, http://nytlive.nytimes.com/womenintheworld/2015/04/24/

in-parts-of-the-us-maternal-death-rates-are-on-par-with-sub-saharan-africa (accessed 21 October 2016).

5. Ian L. Marpuri, "Researchers explore genomic data privacy and risk," National Human Genome Research Institute 8 April 2013, https://www.genome.gov/27553487/researchers-explore-genomic-data-privacy-and-risk (accessed 21 October 2016).

6. The Genetic Information Nondiscrimination Act of 2008, U.S. Equal Employment Opportunity Commission 21 May 2008, https://www.eeoc.gov/laws/statutes/gina.cfm (accessed 21 October 2016).

PART THREE

CHAPTER EIGHT

1. "Planet Money," National Public Radio 8 May 2015, http://www.npr.org/templates/transcript/transcript.php?storyId=405270046 (accessed 21 October 2016).

2. The Verge, "The 2015 DARPA Robotics Challenge Finals," https://www.youtube.com/watch?v=8P9geWwi9e0 (accessed 21 October 2016).

3. Richard Lawler, "Google DeepMind AI wins final Go match for 4–1 series win," *Engadget* 14 March 2016, https://www.engadget.com/2016/03/14/the-final-lee-sedol-vs-alphago-match-is-about-to-start (accessed 21 October 2016).

4. Wan He, Daniel Goodkind, and Paul Kowal, U.S. Census Bureau, *An Aging World: 2015*, International Population Reports P95/16-1, Washington, D.C.: U.S. Government Publishing Office, 2016, http://www.census.gov/content/dam/Census/library/publications/2016/demo/p95-16-1.pdf (accessed 21 October 2016).

5. U.N. Department of Economic and Social Affairs Population Division, *World Population Prospects: The 2015 Revision,* New York: United Nations, 2015, https://esa.un.org/unpd/wpp/Publications/Files/WPP2015_Volume-I_Comprehensive-Tables.pdf (accessed 21 October 2016).

6. Stuart Russell, Nils J. Nilson, Barbara J. Grosz, et al., "Autonomous weapons: An open letter from AI and robotics researchers," Future of Life Institute, http://futureoflife.org/open-letter-autonomous-weapons (accessed 21 October 2016).

7. AJung Moon, "Machine Agency," Roboethics info Database 22 April 2012, http://www.amoon.ca/Roboethics/wiki/the-open-roboethics-initiative/machine-agency.

8. Jason Kravarik and Sara Sidner, "The Dallas shootout, in the eyes of police," CNN 15 July 2016, https://en.wikipedia.org/wiki/2016_shooting_of_Dallas_police_officers (accessed 21 October 2016).

9. Erik Brynjolfsson and Andrew McAfee, *The Second Machine Age: Work, Progress, and Prosperity in a Time of Brilliant Technologies* (rev.), W.W. Norton, 2016, http://books.wwnorton.com/books/The-Second-Machine-Age (accessed 21 October 2016).

10. Michael A. Osborne and Carl Benedikt Frey, *The Future of Employment: How Susceptible Are Jobs to Computerisation?*, Oxford: University of Oxford, 2013, http://futureoflife.org/data/PDF/michael_osborne.pdf (accessed 21 October 2016).

11. James Manyika, Michael Chui, and Mehdi Miremadi, "These are the jobs least likely to go to robots," *Fortune* 11 July 2006, http://fortune.com/2016/07/11/skills-gap-automation.

12. Timothy J. Seppela, "Google is working on a kill switch to prevent an AI uprising," *Engadget* 3 June 2016, https://www.engadget.com/2016/06/03/google-ai-killswitch/ (accessed 21 October 2016).

CHAPTER NINE

1. Dan Kloeffler and Alexis Shaw, "Dick Cheney feared assassination via medical device hacking: 'I was aware of the danger,'" ABC News 19 October 2013, http://abcnews.go.com/US/vice-president-dick-cheney-feared-pacemaker-hacking/story?id=20621434 (accessed 21 October 2016).

2. Kim Zetter, "An unprecedented look at Stuxnet, the world's first digital weapon," *WIRED* 3 November 2014, https://www.wired.com/2014/11/countdown-to-zero-day-stuxnet (accessed 21 October 2016)

3. "What happened," U.S. Office of Personnel Management (undated), https://www.opm.gov/cybersecurity/cybersecurity-incidents (accessed 21 October 2016).

4. Casey Newton, "The mind-bending messiness of the Ashley Madison data dump," the *Verge* 19 August 2015, http://www.theverge.com/2015/8/19/9178855/ashley-madison-data-breach-implications (accessed 21 October 2016).

5. Mat Honan, "How Apple and Amazon security flaws led to my epic hacking," *WIRED* 6 August 2012, https://www.wired.com/2012/08/apple-amazon-mat-honan-hacking (accessed 21 October 2016).

6. Kevin Kelley, *The Inevitable,* Viking: New York, 2016.

CHAPTER TEN

1. Jonathan Vanian, "7-Eleven Just Used a Drone to Deliver a Chicken Sandwich and Slurpees," *Fortune* 22 July 2016, http://fortune.com/2016/07/22/7-eleven-drone-flirtey-slurpee (accessed 21 October 2016).

2. Mary Meeker, "Internet Trends 2015—Code Conference," Kleiner Perkins Caulfield & Byers, http://www.kpcb.com/blog/2015-internet-trends.

3. Chris Anderson, "How I accidentally kickstarted the domestic drone boom," *WIRED* 22 June 2012, http://www.wired.com/2012/06/ff_drones (accessed 21 October 2016).

4. "Malawi tests first unmanned aerial vehicle flights for HIV early infant diagnosis," UNICEF 14 March 2016, http://www.unicef.org/media/media_90462.html (accessed 21 October 2016).

5. Jonathan Vanian, "Drone makes first legal doorstep delivery in milestone flight," *Fortune* 17 July 2015, http://fortune.com/2015/07/17/faa-drone-delivery-amazon (accessed 21 October 2016).

6. Nicole Comstock, "Cal fire air tankers grounded due to drone," Fox40 25 June 2015, http://fox40.com/2015/06/25/cal-fire-air-tankers-grounded-due-to-drone (accessed 21 October 2016).

7. Kristina Davis, "Two plead guilty in border drug smuggling by drone," *Los Angeles Times* 13 August 2015, http://www.latimes.com/local/california/la-me-drone-drugs-20150813-story.html.

8. Victoria Bekiempis, "Father of man who built gun-shooting 'drone' says don't panic," *Newsweek* 21 July 2015, http://www.Newsweek.com/gun-shooting-drone-Newsweek-talks-inventors-dad-355723 (accessed 21 October 2016).

9. "ISIS booby-trapped drone kills troops in Iraq, officials say," *Guardian* 12 October 2016, https://www.theguardian.com/world/2016/oct/12/exploding-drone-sent-by-isis-allies-kills-and-wounds-troops-in-iraq-report (accessed 21 October 2016).

10. "Unmanned Aircraft Systems," Federal Aviation Authority 29 August 2016, https://www.faa.gov/uas/ (accessed 23 October 2016).

11. "Current unmanned aircraft state law landscape," National Conference of State Legislatures 7 October 2016, http://www.ncsl.org/research/transportation/current-unmanned-aircraft-state-law-landscape.aspx (accessed 21 October 2016).

CHAPTER ELEVEN

1. Diana W. Bianchi, R. Lamar Parker, Jeffrey Wentworth, et al., "DNA sequencing versus standard aneuploidy screening," *New England Journal of Medicine* 2014;370:799–808, http://www.nejm.org/doi/full/10.1056/NEJMoa1311037 (accessed 21 October 2016).

2. Jessica X. Chong, Kati J. Buckingham, Shalini N. Jhangiani, et al., "The genetic basis of Mendelian phenotypes: Discoveries, challenges, and opportunities," *American Journal of Human Genetics* 2015;97(2):199–215, http://www.cell.com/ajhg/abstract/S0002-9297%2815% 2900245-1 (accessed 21 October 2016).

3. Aleksandar D. Kostic, Dirk Gevers, Heli Siljander, et al., "The Dynamics of the Human Infant Gut Microbiome in Development and in Progression toward Type 1 Diabetes," *Cell Host & Microbe* 2015;17(2):260–273, http://dx.doi.org/10.1016/j.chom.2015.01.001 (accessed 21 October 2016).

4. David L. Suskind, Mitchell J. Brittnacher, Ghassan Wahbeh, et al., "Fecal microbial transplant effect on clinical outcomes and fecal microbiome in active Crohn's disease," *Inflammatory Bowel Diseases* 2015;21(3):556–563, http://journals.lww.com/ibdjournal/Fulltext/2015/03000/Fecal_Microbial_Transplant_Effect_on_Clinical.7.aspx (accessed 21 October 2016).

5. Kate Lunau, "Scientists are now trying fecal transplants on kids," *Motherboard* 16 September 2016, http://motherboard.vice.com/read/fecal-transplants-kids-ibd-crohns-colitis-mcmaster-clinical-trial (accessed 21 October 2016).

6. Richard J. Turnbaugh, Vanessa K. Ridaura, Jeremiah J. Faith, et al., "The effect of diet on the human gut microbiome: A metagenomic analysis in humanized gnotobiotic mice," *Science Translational Medicine* 2009;1(6):6ra14, https://www.ncbi.nlm.nih.gov/pmc/articles/PMC2894525 (accessed 21 October 2016).

7. "Gen9 announces next generation of the BioFab® DNA synthesis platform," Gen9 21 March 2016, https://blog.gen9bio.com/about-us/news-events/press-releases/gen9-announces

-next-generation-biofab-dna-synthesis-platform (accessed 21 October 2016).

8. Yuyu Niu, Bin Shen, Yiqiang Cui, et al., "Generation of gene-modified cynomolgus monkey via Cas9/RNA-mediated gene targeting in one-cell embryos," *Cell* 2014;156(4):836–843, http://www.cell.com/cell/abstract/S0092-8674(14)00079-8 (accessed 21 October 2016).

9. David Cyranoski and Sara Reardon, "Chinese scientists genetically modify human embryos," *Nature* 22 April 2015, http://www.nature.com/news/chinese-scientists-genetically-modify-human-embryos-1.17378 (accessed 21 October 2016).

10. Ewen Callaway, "Second Chinese team reports gene editing in human embryos," *Nature* 8 April 2016, http://www.nature.com/news/second-chinese-team-reports-gene-editing-in-human-embryos-1.19718 (accessed 21 October 2016).

11. "Controversy over genetically altered mosquitos," *Science Daily* 6 December 2012, https://www.sciencedaily.com/videos/521327.htm (accessed 21 October 2016). Joseph Curtis, "Are scientists to blame for Zika virus? Researchers released genetically modified mosquitos into Brazil three years ago," *Daily Mail* Australia 1 February 2016, http://www.dailymail.co.uk/news/article-3425381/Are-scientists-blame-Zika-virus-Researchers-released-genetically-modified-mosquitos-Brazil-three-years-ago.html. Deanna Ferrante, "Florida residents protest release of genetically modified mosquitos to fight Zika virus," *Orlando Weekly* 22 April 2016, http://www.orlandoweekly.com/Blogs/archives/2016/04/22/florida-residents-protest-release-of-genetically-modified-mosquitos-to-fight-zika-virus.

12. Andrew Pollack, "Jennifer Doudna, a pioneer who helped simplify gene editing," *New York Times* 11 May 2015, http://www.nytimes.com/2015/05/12/science/jennifer-doudna-crispr-cas9-genetic-engineering.html (accessed 21 October 2016).

13. Vivek Wadhwa, "Why there's an urgent need for a moratorium on gene editing," *Washington Post* 8 September 2015, https://www.washingtonpost.com/news/innovations/wp/2015/09/08/why-theres-an-urgent-need-for-a-moratorium-on-gene-editing (accessed 21 October 2016).

14. "International summit on human gene editing, December 1–3 2015," Innovative Genomics Initiative (undated), https://innovative

genomics.org/international-summit-on-human-gene-editing (accessed 21 October 2016).

15. Miles Donovan, "Hacking the President's DNA," *Atlantic* November 2012, http://www.theatlantic.com/magazine/archive/2012/11/hacking-the-presidents-dna/309147 (accessed 21 October 2016).

16. The White House, "FACT SHEET: Announcing the National Microbiome Initiative," The White House 13 May 2016, https://www.whitehouse.gov/the-press-office/2016/05/12/fact-sheet-announcing-national-microbiome-initiative (accessed 21 October 2016).

PART FOUR

CHAPTER TWELVE

1. "If I Built a Car: By Chris Van Dusen," Penguin Random House 14 June 2007, http://www.penguinrandomhouse.com/books/293311/if-i-built-a-car-by-chris-van-dusen/9780142408254 (accessed 21 October 2016).

2. Erin Stepp, "Three-quarters of Americans 'afraid' to ride in a self-driving vehicle," American Automobile Association 1 March 2016, http://newsroom.aaa.com/2016/03/three-quarters-of-americans-afraid-to-ride-in-a-self-driving-vehicle (accessed 21 October 2016).

3. Fred Lambert, "Understanding the fatal Tesla accident on autopilot and the NHTSA probe," *Electrek* 1 July 2016, https://electrek.co/2016/07/01/understanding-fatal-tesla-accident-autopilot-nhtsa-probe (accessed 21 October 2016).

4. Lawrence D. Burns, William C. Jordan, and Bonnie A. Scarborough, *Transforming Personal Mobility* (rev.), New York, NY: The Earth Institute, Columbia University, 2013, http://sustainablemobility.ei.columbia.edu/files/2012/12/Transforming-Personal-Mobility-Jan-27-20132.pdf (accessed 21 October 2016).

5. Paul Stenquist, "In self-driving cars, a potential lifeline for the disabled," *New York Times* (New York edition) 9 November 2014:AU2, http://www.nytimes.com/2014/11/09/automobiles/in-self-driving-cars-a-potential-lifeline-for-the-disabled.html (accessed 21 October 2016).

6. J.R. Treat, N.S. Tumbas, S.T. McDonald, et al., *Tri-Level Study of the Causes of Traffic Accidents: Final Report*, volume II: Special Analyses, Bloomington, Indiana: Institute for Research in Public Safety, 1979, http://ntl.bts.gov/lib/47000/47200/47286/Tri-level_

study_ofrom_the_causes_of_traffic_accidents_vol__II.pdf (accessed 21 October 2016).

7. "Road traffic deaths," World Health Organization 2015, http://www.who.int/gho/road_safety/mortality/en (accessed 23 October 2016).

8. Insurance Institute for Highway Safety's Highway Loss Data Institute, General Statistics 2004, http://www.iihs.org/iihs/topics/t/general-statistics/fatalityfacts/state-by-state-overview (accessed 21 October 2016).

9. World Health Organization, Road Safety, http://gamapserver.who.int/gho/interactive_charts/road_safety/road_traffic_deaths2/atlas.html (accessed 21 October 2016).

10. Alyssa Abkowitz, "Baidu plans to mass produce autonomous cars in five years," *Wall Street Journal* 2 June 2016, http://www.wsj.com/articles/baidu-plans-to-mass-produce-autonomous-cars-in-five-years-1464924067 (accessed 21 October 2016).

11. Annabelle Liang and Dee-Ann Durbin, "World's first self-driving taxis debut in Singapore," The Big Story 25 August 2016, http://bigstory.ap.org/article/615568b7668b452bbc8d2e2f3e5148e6/worlds-first-self-driving-taxis-debut-singapore (accessed 21 October 2016).

12. "Reports, trends & statistics," American Trucking Associations (undated), http://www.trucking.org/News_and_Information_Reports_Industry_Data.aspx (accessed 21 October 2016).

13. Taemie Kim and Pamela Hinds, "Who Should I Blame? Effects of Autonomy and Transparency on Attributions in Human–Robot Interaction" (in: *RO-MAN 2006—The 15th IEEE International Symposium on Robot and Human Interactive Communication*, Cambridge, Massachusetts: M.I.T., 2006), M.I.T. (undated), http://alumni.media.mit.edu/~taemie/papers/200609_ROMAN_TKim.pdf (accessed 21 October 2016).

14. Kirsten Korosec, "Elon Musk says Tesla vehicles will drive themselves in two years," *Fortune* 21 December 2015, http://fortune.com/2015/12/21/elon-musk-interview (accessed 21 October 2016).

15. Max Chafkin, "Uber's first self-driving fleet arrives in Pittsburgh this month," *Bloomberg* 18 August 2016, http://www.bloomberg.com/news/features/2016-08-18/uber-s-first-self-driving-fleet-arrives-in-pittsburgh-this-month-is06r7on (accessed 23 October 2016).

CHAPTER THIRTEEN

1. Generali (undated) http://www.generali.es/seguros-particulares/auto-pago-como-conduzco (accessed 21 October 2016).

2. James Manyika, Michael Chui, Peter Bisson, et al., *The Internet of Things: Mapping the Value beyond the Hype*, McKinsey 2015, http://www.mckinsey.com/business-functions/digital-mckinsey/our-insights/the-internet-of-things-the-value-of-digitizing-the-physical-world (accessed 21 October 2016).

3. Hayley Tsukayama, "VTech says 6.4 million children profiles were caught up in its data breach," *Washington Post* 1 December 2015, https://www.washingtonpost.com/news/the-switch/wp/2015/12/01/vtech-says-6-4-million-children-were-caught-up-in-its-data-breach (accessed 21 October 2016).

4. Lorenzo Franceschi-Bicchierai, "One of the largest hacks yet exposes data on hundreds of thousands of kids," *Motherboard* 27 November 2015, http://motherboard.vice.com/read/one-of-the-largest-hacks-yet-exposes-data-on-hundreds-of-thousands-of-kids (accessed 21 October 2016).

5. Lorenzo Franceschi-Bicchierai, "Hacker obtained children's headshots and chatlogs from toymaker VTech," *Motherboard* 30 November 2015, http://motherboard.vice.com/read/hacker-obtained-childrens-headshots-and-chatlogs-from-toymaker-vtech (accessed 21 October 2016).

6. Andrea Peterson, "Hello (hackable) Barbie," *Washington Post* 4 December 2015, https://www.washingtonpost.com/news/the-switch/wp/2015/12/04/hello-hackable-barbie (accessed 21 October 2016).

7. "FAQ about cyber attack on VTech Learning Lodge," VTech 8 August 2016, https://www.vtech.com/en/press_release/2015/faq-about-data-breach-on-vtech-learning-lodge (accessed 21 October 2016).

8. PwC, *Managing Cyber Risks in an Interconnected World: Key findings from The Global State of Information Security® Survey 2015*, PwC 2014, http://www.pwc.com/gx/en/consulting-services/information-security-survey/assets/the-global-state-of-information-security-survey-2015.pdf (accessed 21 October 2016).

9. "Equipment Authorization Approval Guide," Federal Communications Commission 21 October 2015, https://www.fcc.gov/engineering-technology/laboratory-division/general/equipment-authorization (accessed 21 October 2016).

CHAPTER FOURTEEN

1. Rob Stein, "Baby thrives once 3-D-printed windpipe helps him breathe," NPR 23 December 2014, http://www.npr.org/sections/health-shots/2014/12/23/370381866/baby-thrives-once-3D-printed-windpipe-helps-him-breathe (accessed 21 October 2016).

2. NPR 17 March 2014, http://www.npr.org/sections/health-shots/2014/03/17/289042381/doctors-use-3-d-printing-to-help-a-baby-breathe.

3. Elizabeth Svoboda, "'Watch me walk,'" *Saturday Evening Post* March–April 2012;284(2):20–25, http://www.saturdayeveningpost.com/2012/03/14/in-the-magazine/health-in-the-magazine/watch-walk.html (accessed 21 October 2016).

4. Catherine de Lange, "Engineered vaginas grown in women for the first time," *New Scientist* 10 April 2014, https://www.newscientist.com/article/dn25399-engineered-vaginas-grown-in-women-for-the-first-time (accessed 21 October 2016).

CHAPTER FIFTEEN

1. "Water, sanitation and hygiene links to health," World Health Organization November 2004, http://www.who.int/water_sanitation_health/publications/facts2004/en (accessed 21 October 2016).

2. "A task of terawatts" (editorial), *Nature* 14 August 2008;454:805, http://www.nature.com/nature/journal/v454/n7206/full/454805a.html (accessed 21 October 2016).

3. "Grupo Rotoplas announces agreement to acquire minority stake in the Advanced Innovation Centre (AIC)," PR Newswire 9 March 2016, http://www.prnewswire.com/news-releases/grupo-rotoplas-announces-agreement-to-acquire-minority-stake-in-the-advanced-innovation-center-aic-300233881.html (accessed 21 October 2016).

4. Ramez Naam, *The Infinite Resource: The Power of Ideas on a Finite Planet,* Hanover and London: University Press of New England, 2013.

5. Katie Fehrenbacher, "A jaw-dropping world record solar price was just bid in Abu Dhabi," *Fortune* 19 September 2016, http://for

tune.com/2016/09/19/world-record-solar-price-abu-dhabi (accessed 21 October 2016).

6. D. M. Chapin, C. S. Fuller, and G. L. Pearson, "A new silicon *p–n* junction photocell for converting solar radiation into electrical power," *Journal of Applied Physics* May 1954;25:676–677, http://scitation.aip.org/content/aip/journal/jap/25/5/10.1063/1.1721711 (accessed 21 October 2016).

7. Tom Randall, "Wind and solar are crushing fossil fuels," *Bloomberg* (6 April 2016), http://www.bloomberg.com/news/articles/2016-04-06/wind-and-solar-are-crushing-fossil-fuels (accessed 21 October 2016).

8. Seb Henbest, Elena Giannakopoulou, Ethan Zindler, et al., *New Energy Outlook 2016: Powering a Changing World*, Bloomberg New Energy Finance 2016, https://www.bloomberg.com/company/new-energy-outlook (accessed 21 October 2016).

9. Björn Nykvist and Måns Nilsson, "Rapidly falling costs of battery packs for electric vehicles," *Nature Climate Change* 23 March 2015;5:329–332, http://www.nature.com/nclimate/journal/v5/n4/full/nclimate2564.html (accessed 21 October 2016).

10. "The leapfrog continent," *Economist* 6 June 2015, http://www.economist.com/news/middle-east-and-africa/21653618-falling-cost-renewable-energy-may-allow-africa-bypass (accessed 21 October 2016).

11. *Scaling up access to electricity: The case of Lighting Africa,* Live Wire 2014/20, World Bank, http://documents.worldbank.org/curated/en/804081468200339840/pdf/88701-REPF-BRI-PUBLIC-Box385194B-ADD-SERIES-Live-wire-knowledge-note-series-LW20-New-a-OKR.pdf (accessed 21 October 2016).

12. "Lighting the way," *Economist* 1 September 2012, http://www.economist.com/node/21560983 (accessed 21 October 2016).

13. Sudha Ramachandran, "Water wars: China, India and the great dam rush," *Diplomat* 3 April 2015, http://thediplomat.com/2015/04/water-wars-china-india-and-the-great-dam-rush (accessed 21 October 2016).

14. Kotler, *Abundance.*

ACKNOWLEDGMENTS

My deepest debts and gratitude are always to my family. Tavinder, my wife, is the rock of my life and my biggest supporter. My son Tarun, an expert in his own right on many of these topics, served as my reality check and sounding board. My elder son, Vineet, is my guru on health, work–life balance, and many other things.

I want to thank my friend Alex Salkever, who helped me write this book and worked extremely hard on it. He is the person who got me into writing, about ten years ago, while he was an editor at *BusinessWeek*. He coached me on how to write columns for the business press and then on how to write books such as this. And then there is John P. Harvey, my childhood buddy from Canberra, Australia, and the editor of this book. He played a crucial role, keeping us in line and making crucial suggestions to shape and clarify the narrative.

Most of all, I want to thank the tens of thousands of people I have met or conversed with over the past decade of my journey. This book wouldn't have been possible without all the knowledge, wisdom, and insights you have bestowed upon me. Many thanks.

Last, I'd like to thank the people who rarely get thanks: my book agent, Kathleen Anderson, and my publishers, Neal Mallett and Jeevan Sivasubramaniam. They encouraged Alex and me to redraft the book, crystallize our thoughts, and produce something that I hope has the potential to help you make a difference in our world.

INDEX

ABOUT THE AUTHORS

VIVEK WADHWA is a Distinguished Fellow at Carnegie Mellon University's College of Engineering and a Director of Research at Duke University's Pratt School of Engineering. He is a globally syndicated columnist for the *Washington Post* and author of *The Immigrant Exodus: Why America Is Losing the Global Race to Capture Entrepreneurial Talent*, which was named by the *Economist* as a Book of the Year of 2012, and of *Innovating Women: The Changing Face of Technology*, which documents the struggles and triumphs of women. Wadhwa has held appointments at Stanford Law School, Harvard Law School, and Emory University, and is an adjunct faculty member at Singularity University.

Wadhwa is based in Silicon Valley and researches exponentially advancing technologies that are soon going to change our world. These advances—in fields such as robotics, artificial intelligence, computing, synthetic biology, 3-D printing, medicine, and nanomaterials—are making it possible for small teams to do what was once possible only for governments and large corporations to do: solve the grand challenges in education, water, food, shelter, health, and security. They will also disrupt industries and create many new policy, law, and ethics issues.

In 2012, the U.S. Government awarded Wadhwa distinguished recognition as an Outstanding American by Choice, for his "commitment to this country and to the common civic values that unite us as Americans." He was also named one of the world's Top 100 Global Thinkers by *Foreign Policy* magazine in that year. In June 2013, he was on *TIME* magazine's list of Tech 40, one of forty of the most influential minds in tech. And in September 2015, he was second on a list of "ten men worth emulating" in the *Financial Times*.

Wadhwa teaches subjects such as technology, industry disruption, entrepreneurship, and public policy; researches the policy, law, and ethics issues of exponential technologies; helps prepare students for the real world; and advises several governments. In addition to being a columnist for the *Washington Post*, he is a contributor to *VentureBeat*, *Huffington Post*, LinkedIn's *Influencers* blog, and the American Society of Engineering Education's *Prism* magazine. Prior to joining academia in 2005, Wadhwa founded two software companies.

ALEX SALKEVER is Vice President of Marketing Communications at Mozilla. He formerly has worked as a CMO and marketing executive at several high-tech startups. Prior to his roles in marketing Alex served as Technology Editor at BusinessWeek.com. He also was the co-author, with Wadhwa, of *The Immigrant Exodus,* which was listed as one of the Business Books of the Year by the Economist in 2012. Alex lives in the Bay Area.

Berrett-Koehler is an independent publisher dedicated to an ambitious mission: *Connecting people and ideas to create a world that works for all.*

We believe that the solutions to the world's problems will come from all of us, working at all levels: in our organizations, in our society, and in our own lives. Our BK Business books help people make their organizations more humane, democratic, diverse, and effective (we don't think there's any contradiction there). Our BK Currents books offer pathways to creating a more just, equitable, and sustainable society. Our BK Life books help people create positive change in their lives and align their personal practices with their aspirations for a better world.

All of our books are designed to bring people seeking positive change together around the ideas that empower them to see and shape the world in a new way.

And we strive to practice what we preach. At the core of our approach is Stewardship, a deep sense of responsibility to administer the company for the benefit of all of our stakeholder groups including authors, customers, employees, investors, service providers, and the communities and environment around us. Everything we do is built around this and our other key values of quality, partnership, inclusion, and sustainability.

This is why we are both a B-Corporation and a California Benefit Corporation—a certification and a for-profit legal status that require us to adhere to the highest standards for corporate, social, and environmental performance.

We are grateful to our readers, authors, and other friends of the company who consider themselves to be part of the BK Community. We hope that you, too, will join us in our mission.

A BK Business Book

We hope you enjoy this BK Business book. BK Business books pioneer new leadership and management practices and socially responsible approaches to business. They are designed to provide you with groundbreaking and practical tools to transform your work and organizations while upholding the triple bottom line of people, planet, and profits. High-five!

To find out more, visit **www.bkconnection.com.**

Connecting people and ideas
to create a world that works for all

Dear Reader,

Thank you for picking up this book and joining our worldwide community of Berrett-Koehler readers. We share ideas that bring positive change into people's lives, organizations, and society.

To welcome you, we'd like to offer you a free e-book. You can pick from among twelve of our bestselling books by entering the promotional code **BKP92E** here: http://www.bkconnection.com/welcome.

When you claim your free e-book, we'll also send you a copy of our e-newsletter, the *BK Communiqué*. Although you're free to unsubscribe, there are many benefits to sticking around. In every issue of our newsletter you'll find

- A free e-book
- Tips from famous authors
- Discounts on spotlight titles
- Hilarious insider publishing news
- A chance to win a prize for answering a riddle

Best of all, our readers tell us, "Your newsletter is the only one I actually read." So claim your gift today, and please stay in touch!

Sincerely,

Charlotte Ashlock
Steward of the BK Website

Questions? Comments? Contact me at bkcommunity@bkpub.com.